Lecture Notes in Economics and Mathematical Systems

412

Bettina Kuon

Two-Person Bargaining Experiments with Incomplete Information

Springer-Verlag

Berlin Heidelberg New York
London Paris Tokyo
Hong Kong Barcelona
Budapest

Author

Bettina Kuon
University of Bonn
Wirtschaftstheorie I
Adenauerallee 24–42
D-53113 Bonn, FRG

ISBN 978-3-540-57920-5   ISBN 978-3-642-48777-4 (eBook)
DOI 10.1007/ 978-3-642-48777-4

Library of Congress Cataloging-in-Publication Data
Kuon, Bettina, 1963- . Two-person bargaining experiments with incomplete information / Bettina Kuon. p. cm. – (Lecture-notes in economics and mathematical
systems; 412) Includes bibliographical references.
1. Game theory. 2. Negotiation–Mathematical models. I. Title. II. Title: 2-person
bargaining experiments with incomplete information. III. Series.
HB144.K83   1994   658.4'0353–dc20   94-9061

Typesetting: Camera ready by author
SPIN: 10135247   42/3140-543210 - Printed on acid-free paper

Acknowledgements

I am greatly indebted to Reinhard Selten for his helpful advice and suggestions. He stimulated my interest in experimental economics and my views have been refined in many inspiring conversations.

It is a pleasure to acknowledge the help received from my colleagues at the Bonn Laboratory of Experimental Economics, especially Joachim Buchta and Abdol-karim Sadrieh. I have greatly benefitted from discussions with John O. Ledyard, Thomas R. Palfrey, and Charles R. Plott.

With gratitude I acknowledge the financial support by the Deutsche Forschungs-gemeinschaft through the Sonderforschungsbereich 303.

Last but most, I would like to acknowledge the encouragement and support of my husband Siegfried, my parents Wilhelm and Helene Rockenbach, and my sister Beatrice.

CONTENTS

PART III: THE STRATEGY EXPERIMENT

CHAPTER 1. INTRODUCTION

Think of the following situation: A project yielding a gross profit of 100 is offered to two firms. The project can only be conducted by a cooperation of the two firms. No firm is able to conduct the project alone. In order to receive the project the firms have to agree on the allocation of the gross profit. Each of both firms has an alternative project it conducts in case the joint project is not realized. The profitability of an allocation of the joint gross profit for a firm depends on the gross profit from its alternative project.

The gross profit from an alternative project can be either 0 (low alternative value) or $0 < \alpha < 100$ (high alternative value). We say a firm with an alternative value of 0 is of *type L* or *weak*, and a firm with an alternative value of α is of *type H* or *strong*. Both firms can be equally likely weak or strong. The type of one firm is independent of the other firm's type. Each firm knows its own type but the other firm's type is unknown. The only information about the other firm's type is that it is weak or strong, both equally likely and independent of the own type. The value of α is common knowledge.

The firms start bargaining about the allocation of the joint gross profit 100. They propose allocations in an alternating order until they either agree or one firm declares that it refrains from cooperating. In the latter case both firms conduct their alternative project. The firms are unable to prove that they are of a certain type. This means that the incompleteness of the information cannot be removed by communication. With each proposal on the allocation of the 100, a 1% discounting of the 100 as well as of the firms' alternative values occurs.

In case of an agreement on the allocation of the 100 the gross profit of a firm is the discounted agreed share of the 100, in case that one firm declares the break off of the bargaining, the gross profit of a firm is its discounted alternative value.

Note, that in a game with $\alpha > 50$ two firms of type H cannot agree on an allocation yielding an individually rational outcome for both firms. The firms, however, do not know whether they face this situation.

The above example illustrates the *two-person bargaining problem with incomplete informa-*

tion which will be investigated in this book. Of course, we shall describe the bargaining problem in a more formal and concise way, but the example already covers all relevant features of the problem. Three different approaches to the problem will be taken: The *game theoretic analysis* (in Part I), the study of the spontaneous behavior of subjects in a *game playing experiment* (in Part II), and the investigation of strategies programmed by highly experienced subjects in a *strategy experiment* (in Part III). The two different experimental approaches allow to study the bargaining behavior which emerges spontaneously in interactive plays of two subjects, and moreover the instructions experienced subjects give to a representative (strategies). The three approaches together provide a vivid picture of theoretical and experimentally observed behavior in the two-person bargaining problem under consideration.

In what follows we shall shortly report the methods and some of the main findings and indicate the link to results known from the literature. A more detailed discussion of the results is presented in Chapter 15.

The first part of the book is concerned with the modelling and the game theoretic analysis of the bargaining problem. The introduction of incomplete information into a bargaining problem complicates its game theoretic analysis tremendously. Therefore, the literature on two-person bargaining models with incomplete information mostly consists of bargaining problems with restricted strategic possibilities which allow to find a unique sequential equilibrium, or on the other hand, of the analysis of strategically richer games which yield a large variety of equilibria. A complete game theoretic analysis of the bargaining game under consideration has not been presented in the literature. Chatterjee and Samuelson (1987-1988) analyze an equivalent buyer-seller problem under incomplete information. They select a sequential equilibrium and show the existence of other equilibria. We shall investigate the set of the Nash equilibrium points in pure strategies of this game. A refinement of the equilibrium concept, to sequential equilibria for example, would be a narrowing from a behavioral point of view, since it is not obvious that subjects in an experiment are guided by the concept of sequential rationality. It seems to be adequate to examine conditions imposed by weak equilibrium requirements.

We shall deduce necessary and sufficient conditions for Nash equilibrium points in pure and finite strategies. The necessary conditions contain the well known "ex ante" individual rationality and incentive compatibility constraints (Myerson, 1978 and Selten, 1975) as well as additional individual rationality and incentive compatibility constraints which emerge from the extensive form of the game. If both types of the opponent use different strategies, the player is able to distinguish between the two types at some stage of the game. In this case a continuation of the game has to be individually rational and incentive compatible for the player. It can be shown that a large variety of pure Nash equilibria exists. Accordingly, the prediction of the bargaining result by game theory is weak.

The second part of the book studies the spontaneous behavior of subjects in a game playing experiment. In a computer laboratory subjects play 16 plays of the bargaining game with anonymous and changing opponents. A total of 30 sessions was conducted, six for each of the α values 30, 45, 55, 60, and 70 (the five treatments). Each session is an independent observation. In each session we can distinguish four levels of experience, which provides the opportunity to investigate the changes of the subjects' behavior with experience.

In the third part of the book we study strategies which are developed by 31 highly experienced subjects. The subjects participated in a strategy experiment over four months. After they gained experience with the two-person bargaining problem in game playing sessions each subject had to develop a complete strategy for the bargaining problem with $\alpha=30$, 45, and 60. In three subsequent tournaments a subject could observe the performance of his strategy and, if necessary, improve it.

The analysis of the data of the game playing experiment mainly emphasizes agreement outcomes, circumstances of break offs, initial demands, revelation behavior of weak players, concession behavior, the bargaining process, and the subject's adaptation to experience. The analysis of the strategy experiment mainly emphasizes the picture of the typical strategy. With the help of the method by Kuon (1993) the typicalness of a strategy and the typicalness of each characteristic of the strategies are determined. In a more rigorous form aspects already detected in the game playing experiment are found again.

It is possible to develop a descriptive theory of the agreement outcomes which combines the subjects' tendency to choose prominent numbers with the strategic aspects emerging from the incomplete information. In games with $\alpha < 50$ experienced subjects of the game playing experiment as well as the strategies of the strategy experiment agree on the equal split of the 100. Already in previous experiments involving incomplete information (inter alia Roth and Murnighan, 1982) it was found that subjects overcome the "unclear" situation of incomplete information by applying familiar social norms, like the equal split. For $\alpha < 50$ the equal split yields individually rational payoffs for both types and can therefore serve as a focal allocation. In contrast, the agreement outcomes are highly asymmetric for $\alpha > 50$. This is especially true for plays of two weak players. These agreement outcomes are similar to those in plays of a weak and a strong player. The asymmetry even rises with experience. This is in accord with findings by Hoggatt et al. (1978) in a similar two-person game with incomplete information. A major reason for the occurrence of asymmetry in the outcomes in plays of two weak players is the *sudden acceptance*. In the situation that both weak players pretend to be strong, it often happens that one weak player suddenly accepts the proposal of the opponent. This can be found in the game playing experiment and, by sharp drops in the acceptance levels, also in the strategy experiment. It can be explained by the fear of a break off by the possibly strong opponent. Indeed, the main reason for a break off by a strong player was that he did not receive an offer yielding an individually rational amount.

The strong players, however, are satisfied with an only small additional gain from cooperation. They face the problem that they cannot prove their strength during the bargaining process, so that their average outcome is smaller than in comparable games with complete information (Kuon and Uhlich, 1993). Most surprisingly, the strong types of the game playing experiment were, on average, not able to receive individually rational average payoffs in games with $\alpha > 50$. The agreement payoff was too low to compensate discount losses which occurred in case of conflict. A non-individually rational payoff contradicts the hypothesis that the subjects play Nash equilibria. In the strategy experiment the strong types show a more rigorous break off behavior in plays with $\alpha > 50$, so that most of the strategies are able to reach average payoffs which are individually rational.

A simple qualitative learning theory can be found for the adaptation of the bargaining behavior from play to play. A subject of the game playing experiment weakens its bargaining behavior after a break off in the previous play and it strengthens its bargaining behavior after an agreement. The adaptation after an agreement is less strong than after a break off.

CHAPTER 2. TWO-PERSON BARGAINING WITH INCOMPLETE INFORMATION

This chapter starts with the presentation of the two-person bargaining game with incomplete information which is investigated in this book. In Section 2.2 this special problem will be related to the literature on bargaining games with incomplete information. Especially, we shall discuss a similar bargaining problem, for which a game theoretic solution is already known.

2.1 THE TWO-PERSON BARGAINING GAME WITH INCOMPLETE INFORMATION

Two players bargain over the allocation of the coalition value 100. The bargaining ends in case of an agreement or if one player breaks off the bargaining. If the players agree on an allocation each player's outcome is his agreed share of the coalition value. If a player breaks off the bargaining each player's outcome is his alternative value, which can be either high or low, both equally likely. The low alternative value is 0 and we say that a player with a low alternative value is *the low type*, *weak* or is *of type L*. We say a player is *the high type*, *strong* or is *of type H* if he has the high alternative value $0 < \alpha < 100$. The value of α is common knowledge. Two independent random draws determine the types of the players before the bargaining starts. Hence, the four possible type combinations (L,L), (L,H), (H,L), and (H,H) have equal probability. Each player knows his own type but is ignorant of the other player's type. The only information about the opponent's type is that it can be either H or L, both with probability ½, independent of the player's own type. The players have no instrument to prove their type or to verify the opponent's type during the bargaining. After the termination of the bargaining the information about the opponent's type is not provided.

The bargaining proceeds in a sequential order. Time is divided into stages. A random draw, in which both players are chosen equally likely, determines the first decider. The first decider will be called the *first mover*, or just *player 1*. The other player will be called *second mover* or *player 2*.

A decider has the following options:
- *propose* an allocation of the coalition value,
- *accept* the last proposal of the opponent (not available for the first decider at the beginning of the game),
- *break off* the bargaining.

A player proposes an allocation of the coalition value by stating a demand $0 \leq x \leq 100$ for his own outcome. In case the opponent accepts this proposal his outcome is $100 - x$. The demand x has to be a multiple of the smallest money unit μ.

The bargaining proceeds in successive steps where the players alternate in deciding. One decision will be called one *step* of the bargaining such that the decision steps of the first mover are the odd steps while the second mover has to decide at all even steps of the bargaining. The bargaining ends if either an acceptance or a break off occurs. There is no limit to the bargaining time or to the number of steps.

If in step $n+1$ a player accepts the allocation which was proposed by the opponent in step n, we say the bargaining ended in *agreement after n steps* (in step n). If a player breaks off in step n, we say the bargaining ended by *break off in step n*.

With each proposal after the initial proposal by the first mover the coalition value as well as the alternative values of both players are discounted by a factor $0 < \delta \leq 1$. Hence, in case of an agreement after n steps the payoff of a player is his agreed outcome multiplied with δ^{n-1}, and in case of break off in step n the payoff of a player is his alternative value multiplied with δ^{n-1}. The payoff of an infinite play is defined as zero for both players.

If necessary, we shall distinguish the cases $\delta = 1$ (*no discounting*) and $0 < \delta < 1$ (*discounting*).

2.2   TWO-PERSON BARGAINING MODELS WITH INCOMPLETE INFORMATION

A fundamental approach to bargaining under incomplete information is due to Harsanyi and Selten (1972). They extended the Nash bargaining solution (Nash, 1950) to the case of incomplete information. The theory of Harsanyi and Selten rests on a set of eight axioms which they suggest a bargaining solution should satisfy, and they show that these axioms single out a unique solution for the two-person bargaining game under incomplete information. The investigation is based on an earlier work by Harsanyi (1967-1968), where he introduces the Bayesian approach to games with incomplete information. This allows a concise modelling of games with incomplete information.

Myerson (1979) proposes the *incentive-feasible bargaining solution* and shows that it coincides with the approach by Harsanyi and Selten, this means it maximizes the *generalized Nash product*. In a later work (Myerson, 1984) he proposes the *neutral bargaining solution*, a new axiomatic approach for a two-person Bayesian bargaining problem with incomplete information.

Rubinstein (1985a,b) chooses a different approach for the solution of bargaining games with incomplete information, the *strategic approach*. If each player has complete information about the other's preferences Rubinstein (1982) shows that the alternating offer game with discounting has a unique subgame perfect equilibrium. The introduction of incomplete information to this game creates the difficulty that the beliefs of the players have to be taken into account, in particular the beliefs off the equilibrium path. Mostly, the consequence is a large variety of sequential equilibria. Rubinstein overcomes this problem by imposing additional requirements on beliefs and on equilibrium behavior. Rubinstein (1985a,b) points out the connection between the choice of believes and the sequential equilibrium of the game.

Besides these fundamental investigations in bargaining games with incomplete information there is a large variety of examinations of special games with incomplete information (see Ausubel and Deneckere, 1989, and Selten, 1982 for overviews). Most of these bargaining games are either very restricted due to their strategic possibilities (in order to achieve unique equilibria) or yield a large variety of equilibrium points. In the latter case most authors prefer to introduce additional requirements in order to single out equilibria.

The two-person bargaining game with incomplete information under consideration is a generalization of the alternating offer game (Rubinstein, 1982) to the case of two-sided incomplete information with the possibility to quit the game in each decision step. Nevertheless, an overview over the set of Nash equilibria for this game is not given in the literature.

However, there are two related studies by Chatterjee and Samuelson (1987-1988) investigating in a buyer-seller-problem which can be translated equivalently into the bargaining game under consideration. Under additional restrictions Chatterjee and Samuelson determine Nash equilibria for this game.

The model considers a market with a seller (possessing one indivisible unit of a good) and a buyer (not possessing the good). They can be of two possible types: a *hard* and a *soft* type, with exogenously given probabilities which are common knowledge. The type of the player is only known to the player and not to the opponent. The type determines the valuation of the seller (buyer) which makes him indifferent between selling (buying) or not. These valuations fulfill $s \leq B < S \leq b$, where s denotes the valuation of a soft seller, S the valuation of a hard seller, b the valuation of a soft buyer, and B denotes the valuation of a hard buyer.

The bargaining proceeds as follows. One player is randomly chosen to be the seller. The seller makes the first offer. If the buyer agrees the trade occurs and the game ends. If the buyer rejects, the payoffs of the buyer and the seller are discounted (with potentially different discount factors) and the buyer is the next proposer. The game proceeds in this alternating offer sequence until either an agreement is reached or one player quits the game (which is possible at any time).

Chatterjee and Samuelson (1987) restrict the offers to be from the set $\{b,S\}$ and find a unique Nash equilibrium, which is also a sequential equilibrium.

The buyer-seller-problem has a close relationship to the two-person bargaining problem with incomplete information investigated here. Choosing the parameters for the valuations and the discount factors in an appropriate way, this game coincides with the bargaining game under consideration. The valuations of the two players are either soft, which means that the player has the alternative value 0, or hard, which means that the player has the alternative value α. From the above relationship of the valuations it follows that α has to be greater than 50. The restriction in the offers, translated to the bargaining problem, means that the players can offer either α or $100-\alpha$. The unique Nash equilibrium, Chatterjee and Samuelson found, prescribes that hard players always offer α, and soft players randomize between offering α and $100-\alpha$. A Nash equilibrium in pure strategies only exists if the probability of a soft opponent is very small. Then it is more profitable to propose $100-\alpha$ immediately. For the example of $\alpha=60$ this probability has to be smaller than .02.

Chatterjee and Samuelson (1987) found a unique Nash equilibrium for the bargaining game under consideration for $\alpha>50$ only under a severe restriction of the admissible offers. Thus, if one considers the bargaining problem for $\alpha=60$, and assumes that the players are allowed to demand either 60 or 40, the equilibrium prescribes that a player with the alternative 60 will always demand 60 and a player with the alternative 0 will mix between demanding 40 and 60 (given an equal probability of the two types).

The restriction to these two offers is extremely narrowing. For the strong player the offer of α is the only one which is individually rational (in the feasible set). Moreover, the author wonders why the strong player does not immediately quit the game. Due to the discounting and the restriction in the offers he cannot improve his payoff by playing the game. Unfortunately, Chatterjee and Samuelson do not discuss this problem, although they explicitly mention that a player can quit the game at any time.

In a later paper Chatterjee and Samuelson (1988) give up the restriction on the offers. Now, a variety of equilibria exists. The authors single out a sequential equilibrium which shares features of the equilibrium of the restricted offer case. The hard agents play the pure strategy of demanding the unique sequential equilibrium demand of the complete information game against a soft opponent (*concealing demand*) and accepting no offer worse than this. The soft agents randomize between imitating the hard agent's offer and demanding a value which reveals the (soft) type. If the opponent makes a concealing offer the player rejects this offer and revises downwards the probability describing the likelihood that the opponent is soft. However, if this probability reaches a critical value, the concealing offer is either accepted or the soft agent makes a revealing offer with probability one. The latter case initiates a quasi subgame of one-sided incomplete information. The uninformed agent makes a series of increasingly favorable offers to the informed agent which are chosen to make the soft informed agent indifferent between accepting and waiting for the next more favorable offer. The soft informed agent randomizes between accepting and rejecting these offers. The informed agents make concealing offers which are rejected by the uninformed agent. The uninformed agent revises downwards the probability describing the likelihood that the opponent is soft. As this probability crosses a critical value the uninformed agent accepts the informed agent's offer.

A very specific prediction like that of Chatterjee and Samuelson has a small chance to fit the data exactly. Therefore it is desirable to obtain an overview over a much broader class of game theoretical predictions. In Chapter 3 an attempt will be made to describe the set of all pure strategy Nash equilibria.

CHAPTER 3. THE NASH EQUILIBRIA IN PURE STRATEGIES

A full game theoretic analysis of the two-person game with incomplete information is not available in the literature. It might be helpful to know something about the set of the equilibrium points of this game in order to make a prediction of the experimental results and the shape of the strategies. Therefore, we shall solve the game for Nash equilibria in pure strategies. A refinement of the equilibrium concept, to sequential equilibria for example, would be a narrowing from a behavioral point of view. It is not obvious that subjects follow the concept of sequential rationality. Since a large variety of possible behavior in equilibrium should be captured by the analysis we shall concentrate on the set of pure Nash equilibria. The restriction to pure strategies was chosen in order to simplify the computations. In this respect it is of interest that in the strategy experiment only pure strategies are observed.

In the game theoretic analysis we restrict the range of feasible demands x to $0 < x < 100$. This means we exclude the extreme demands 0 and 100. This simplifies the analysis since we do not have to consider the low type's indifference between a proposal of 0 and the conflict outcome of 0. For a further discussion see also Section 5.2.

3.1   NECESSARY AND SUFFICIENT CONDITIONS

Myerson (1979) showed that for a bargaining problem with incomplete information, like the one investigated here, the Nash equilibria will always lead to allocations in the set of all incentive compatible payoff vectors which are individually rational. The property of incentive compatibility requires that no player should expect any positive gain from imitating another type when all others stick to the equilibrium. Individual rationality requires that no player of any type expects to do worse than in the conflict outcome.

Besides this "ex ante" individual rationality and incentive compatibility, also described in Selten (1975), partial individual rationality and incentive compatibility constraints emerge as additional necessary conditions for Nash equilibria. These constraints are implied by the extensive form of the game. It may happen that the two types of the opponent play different strategies, and therefore the player can distinguish between both types of the opponent at some stage of the game. If the player then has an incentive to deviate from his strategy, it cannot be an equilibrium strategy.

NOTATION

Let s denote a pure strategy tuple for the two-person bargaining game with incomplete information. A strategy tuple contains four strategies: a strategy for the first mover of type H, a strategy for the first mover of type L, a strategy for the second mover of type H, and a strategy for the second mover of type L. The strategies are complete plans for playing the game. We shall restrict our attention to finite strategies. A *finite strategy* is a strategy which ends the game by break off or acceptance in finite time for every opponent's strategy (finite or infinite) and for both types of the opponent. The restriction to finite strategies ensures strictly positive agreement payoffs.

Consider a fixed pure strategy tuple s. Let p_{ij} denote the probability of an agreement in a play of a first mover of type i and a second mover of type j and let p_{ij}^* denote the complementary probability $1 - p_{ij}$, for $i,j \in \{H,L\}$. Since we restricted the attention to pure strategies, both p_{ij} and p_{ij}^* can only take the values 0 and 1. Let $p = (p_{HH}, p_{HL}, p_{LH}, p_{LL})$ denote the vector of the agreement probabilities.

Let a_{ij} denote the outcome of type i in a play of a first mover of type i and a second mover of type j, in case of agreement ($p_{ij} = 1$). The variable a_{ij} can take all integer values strictly greater than 0, and strictly smaller than 100. Consequently, $100 - a_{ij}$ is the agreement outcome of type j in this play. If $p_{ij} = 0$ (this means that no agreement in a play of a first mover of type i and a second mover of type j occurs) define $a_{ij} = 0$. In this case the types receive their alternative values as outcomes.

Let n_{ij} be the number of steps in which the agreement in a play of a first mover of type i and a second mover of type j is reached, if $p_{ij} = 1$. Let n_{ij}^* be the number of steps in which the conflict in a play of a first mover of type i and a second mover of type j is reached, if $p_{ij} = 0$. The variables n_{ij} and n_{ij}^* are integer variables, strictly greater than 0. Define $n_{ij} = 1$, if $p_{ij} = 0$, and define $n_{ij}^* = 1$, if $p_{ij} = 1$. Accordingly, $t_{ij} = p_{ij} n_{ij} + p_{ij}^* n_{ij}^*$ is the step of the termination of the play of a first mover of type i and a second mover of type j.

Let δ_{ij} denote the discount factor which emerged in a play of a first mover of type i and a second mover of type j, if $p_{ij}=1$. Let δ_{ij}^{*} denote the discount factor that emerges in a play of a first mover i and a second mover j, if $p_{ij}=0$. Due to the structure of the discounting the variables δ_{ij} and δ_{ij}^{*} are real variables of the form $\delta_{ij} = \delta^{n_{ij}-1}$ and $\delta_{ij}^{*} = \delta^{n_{ij}^{*}-1}$. Define $\delta_{ij}=1$, if $p_{ij}=0$, and define $\delta_{ij}^{*}=1$, if $p_{ij}^{*}=0$.

The restriction to finite strategies yields $n_{ij} < \infty$ and $n_{ij}^{*} < \infty$, which implies $\delta_{ij}>0$ and $\delta_{ij}^{*}>0$, for all $i,j \in \{H,L\}$.

We shall refer to the ranges of the parameters which are induced by the problem as the *admissible parameter ranges*.

We shall make a distinction between the outcome and the payoff of a player. The *outcome* denotes the non-discounted result of the player after the termination of the play. This is either the share of the coalition value which a player received in agreement or the alternative value in case of conflict. The *payoff* is the discounted outcome.

The Parameter Space of the Equilibria

As far as outcomes and payoffs of the two-person bargaining problem with incomplete information are concerned, equilibrium points in pure strategies can be described by twelve parameters: the agreement probabilities p_{ij}, the agreement outcomes a_{ij}, and the steps n_{ij} (n_{ij}^{*}), for each of the four type combinations $i,j \in \{H,L\}$. If $\delta<1$, the parameters δ_{ij} (δ_{ij}^{*}) are in a bijective relationship to the n_{ij} (n_{ij}^{*}), and if $\delta=1$ the parameters δ_{ij} (δ_{ij}^{*}) are all equal to 1. Therefore, they do not add further dimensions to the description.

The parameter space spanned up by the twelve parameters will be investigated with respect to the restrictions imposed by the condition that a parameter constellation belongs to an equilibrium point in pure and finite strategies. Necessary conditions will be derived and it will be shown that for each parameter constellation satisfying these conditions a pure and finite equilibrium point with this constellation can be constructed.

THE EXPECTED PAYOFFS

If the play of a first mover of type i and a second mover of type j ends in agreement in step n_{ij}, the following payoffs result for the player of type i and the player of type j:

$\delta_{ij} a_{ij} = \delta^{n_{ij}-1} a_{ij}$ for type i and $\delta_{ij}(100-a_{ij}) = \delta^{n_{ij}-1}(100-a_{ij})$ for type j.

If a high type player is involved in a play which ends in conflict in step n_{ij}^{*}, he receives the payoff $\delta_{ij}^{*}\alpha = \delta^{n_{ij}^{*}-1}\alpha$. If a low type player is involved in a play which ends in conflict in step n_{ij}^{*}, he receives the payoff 0.

The *expected payoffs* of the types from playing the two-person game with incomplete information can now be determined. A player faces with equal probability a weak opponent and a strong opponent.

$$P_H^1 = .5 \cdot (p_{HH}\delta_{HH}a_{HH} + p_{HH}^{*}\delta_{HH}^{*}\alpha) + .5 \cdot (p_{HL}\delta_{HL}a_{HL} + p_{HL}^{*}\delta_{HL}^{*}\alpha) \tag{E1}$$

$$P_H^2 = .5 \cdot (p_{HH}\delta_{HH}(100-a_{HH}) + p_{HH}^{*}\delta_{HH}^{*}\alpha) + .5 \cdot (p_{LH}\delta_{LH}(100-a_{LH}) + p_{LH}^{*}\delta_{LH}^{*}\alpha) \tag{E2}$$

$$P_L^1 = .5 \cdot (p_{LH}\delta_{LH}a_{LH} + p_{LL}\delta_{LL}a_{LL}) \tag{E3}$$

$$P_L^2 = .5 \cdot (p_{HL}\delta_{HL}(100-a_{HL}) + p_{LL}\delta_{LL}(100-a_{LL})) \tag{E4}$$

The superscript of P indicates the mover while the subscript refers to the type.

INDIVIDUAL RATIONALITY CONSTRAINTS

A first mover of type H can guarantee himself a payoff of α by breaking off in the first step, and a second mover of type H can guarantee himself a payoff of $\delta\alpha$ by breaking off in the second step (his first decision step). To break off the second mover has to reject the proposal of the first mover and thereby he causes one discount step.

The *individual rationality constraints* demand that the expected payoff of a player has to be at least the value the player can ensure for himself. If a strategy tuple would not satisfy this restriction, it cannot be in equilibrium.

$$P_H^1 \geq \alpha \qquad\qquad\qquad\qquad (R1)$$

$$P_H^2 \geq \delta\alpha \qquad\qquad\qquad\qquad (R2)$$

We shall neglect the individual rationality constraints for the type L players since they are always fulfilled. Notice, that due to the extensive form of the game the two high types have different bounds for the expected payoffs. For the second mover it is the once discounted alternative value, since this is the maximal value he can guarantee for himself.

INCENTIVE CONSTRAINTS

Revealing the true type should be optimal for each player. This means that a player of type L should not have a higher expected payoff by imitating a player of type H than by playing according his own strategy. Analogously, this has to be true for a player of type H. *Imitating* a type means playing according to the strategy of this type. If a strategy tuple would not satisfy these restrictions, it cannot be in equilibrium.

The property of incentive compatibility is expressed by the following *incentive constraints*.

$$P_H^1 \geq .5 \cdot (p_{LH}\delta_{LH}a_{LH} + \overset{\bullet}{p}_{LH}\overset{\bullet}{\delta}_{LH}\alpha) + .5 \cdot (p_{LL}\delta_{LL}a_{LL} + \overset{\bullet}{p}_{LL}\overset{\bullet}{\delta}_{LL}\alpha) \qquad (I1)$$

$$P_H^2 \geq .5 \cdot (p_{HL}\delta_{HL}(100-a_{HL}) + \overset{\bullet}{p}_{HL}\overset{\bullet}{\delta}_{HL}\alpha) + .5 \cdot (p_{LL}\delta_{LL}(100-a_{LL}) + \overset{\bullet}{p}_{LL}\overset{\bullet}{\delta}_{LL}\alpha) \qquad (I2)$$

$$P_L^1 \geq .5 \cdot (p_{HH}\delta_{HH}a_{HH} + p_{HL}\delta_{HL}a_{HL}) \qquad (I3)$$

$$P_L^2 \geq .5 \cdot (p_{HH}\delta_{HH}(100-a_{HH}) + p_{LH}\delta_{LH}(100-a_{LH})) \qquad (I4)$$

The individual rationality and incentive constraints stated above are both ex ante. Given the information at the beginning of the game, a player should expect to gain at least as much as he can guarantee himself, and he should not have an incentive to imitate the other type. Due to the extensive form of the game individual rationality and incentive constraints may also play a role during the game.

PARTIAL INDIVIDUAL RATIONALITY CONSTRAINTS

Whenever a player of type H has an agreement, the agreement payoff should be at least as profitable as a unilateral break off by the strong player at this time. Suppose, for example, H_1 and L_2 are having an agreement in step n_{HL} and the agreement allocation is proposed by H_1. Then $\delta_{HL} a_{HL} \geq \delta_{HL} \alpha$ has to be fulfilled, which means that H_1 is only "willing to demand" a_{HL} if his payoff in case of acceptance is at least as profitable as a break off by H_1 in step n_{HL}. In case that the agreement allocation is proposed by type L_2, $\delta_{HL} a_{HL} \geq \delta \delta_{HL} \alpha$ has to be fulfilled. In this case a break off by H_1 can only occur after he rejected the proposal, such that an additional discount step occurs. In general, the *partial individual rationality constraints* are as follows.

$$
p_{Hj} = 1 \Rightarrow \begin{cases} a_{Hj} \geq \alpha, & \text{if } n_{Hj} \text{ odd} \\[2mm] a_{Hj} \geq \delta\alpha, & \text{if } n_{Hj} \text{ even} \end{cases} \quad j \in \{H, L\}
$$

$$\text{(PR1)}$$

$$
p_{iH} = 1 \Rightarrow \begin{cases} 100 - a_{iH} \geq \alpha, & \text{if } n_{iH} \text{ even} \\[2mm] 100 - a_{iH} \geq \delta\alpha, & \text{if } n_{iH} \text{ odd} \end{cases} \quad i \in \{H, L\}
$$

$$\text{(PR2)}$$

Obviously, it is not necessary to state constraints for the low type since they would be fulfilled trivially. A set of strategies not satisfying the partial individual rationality constraints cannot be in equilibrium since a unilateral deviation would be profitable for the high type.

CONTINUATION CONSTRAINTS

Suppose $p_{HH} = 0$ and $p_{HL} = 1$. This means that H_1 has a conflict with H_2 in step n_{HH}^* and an agreement with L_2 in step n_{HL}. In particular, this means that both types of player 2 cannot play according to the same strategy. Therefore, there must be a step such that the two strategies separate. An upper bound for this step is $n_1 = \min\{n_{HH}^*, n_{HL}\}$. If the game is not terminated in step n_1 (by acceptance or break off), type H_1 is aware of the type of the opponent. Loosely speaking, the *continuation individual rationality* and the *continuation incentive* *constraints* state that the continuation of the game should be individually rational and incentive compatible for H_1 under this additional information. Additional information about the opponent's type is not only obtained if the player reaches a conflict with one type and an agreement with the other type of the opponent, it also is obtained if the player reaches an agreement (or a conflict) with both types, but at different steps. The conditions under which continuation constraints necessarily have to occur are presented in what follows.

Remember, that $t_{ij} = p_{ij} n_{ij} + p_{ij}^* n_{ij}^*$ is the step of the termination of a play of a first mover of type i and a second mover of type j, $i, j \in \{H, L\}$. In case of agreement $t_{ij} = n_{ij}$ and in case of conflict $t_{ij} = n_{ij}^*$.

Condition F(i), $i \in \{H, L\}$

$p_{iL} = 1$ and $p_{iH} = 0$ and (n_{iH}^* even if $t_{iH} < t_{iL}$) $\hspace{2cm}$ (F1)

$p_{iL} = 0$ and $p_{iH} = 1$ and (n_{iL}^* even if $t_{iL} < t_{iH}$) $\hspace{2cm}$ (F2)

$p_{iL} = 0$ and $p_{iH} = 0$ and (n_{iL}^* even if $t_{iL} < t_{iH}$) $\hspace{2cm}$ (F3)

$p_{iL} = 0$ and $p_{iH} = 0$ and (n_{iH}^* even if $t_{iH} < t_{iL}$) $\hspace{2cm}$ (F4)

$p_{iL} = 1$ and $p_{iH} = 1$ and $t_{iH} \neq t_{iL}$ $\hspace{4cm}$ (F5)

Condition F(i) is fulfilled if one of the conditions (F1) to (F5) is fulfilled.

Condition S(j), $j \in \{H, L\}$

$p_{Hj} = 1$ and $p_{Lj} = 0$ and (n_{Lj}^* odd if $t_{Lj} < t_{Hj}$) $\hspace{2cm}$ (S1)

$p_{Hj} = 0$ and $p_{Lj} = 1$ and (n_{Hj}^* odd if $t_{Hj} < t_{Lj}$) $\hspace{2cm}$ (S2)

$p_{Hj} = 0$ and $p_{Lj} = 0$ and (n_{Hj}^* odd if $t_{Hj} < t_{Lj}$) $\hspace{2cm}$ (S3)

$p_{Hj} = 0$ and $p_{Lj} = 0$ and (n_{Lj}^* odd if $t_{Lj} < t_{Hj}$) $\hspace{2cm}$ (S4)

$p_{Hj} = 1$ and $p_{Lj} = 1$ and $t_{Hj} \neq t_{Lj}$ $\hspace{4cm}$ (S5)

Condition S(j) is fulfilled if one of the conditions (S1) to (S5) is fulfilled.

If condition F(H) is fulfilled then there exists a positive probability such that the game is not terminated (by break off or agreement) in step $\min\{t_{HH},t_{HL}\}$, since in case of a conflict it is not initiated by the high type first mover. If the game is not terminated in step $\min\{t_{HH},t_{HL}\}$, type H of player 1 can *distinguish* between both types of player 2 in the continuation of the game. The conditions S(j) are analogous conditions for the second movers.

It is obvious, that under the condition F(i) the inequality $t_{iH} \neq t_{iL}$ holds, and that $t_{Hj} \neq t_{Lj}$ has to be satisfied under the condition S(j), $i,j \in \{H,L\}$.

CONTINUATION INDIVIDUAL RATIONALITY CONSTRAINTS

The *continuation individual rationality constraints* express that in case a player can distinguish between the two types of his opponent, this means conditions F(i) or S(j) are fulfilled, the continuation of the game should be at least as profitable as a unilateral break off by the player. Like the previous individual rationality constraints, these restrictions only have to be stated for the high types since they are always true for the low types. If the unilateral break off would be more profitable than the continuation of the play, the set of strategies cannot be in equilibrium.

$$p_{HH}\delta_{HH}a_{HH} + \overset{\bullet}{p}_{HH}\overset{\bullet}{\delta}_{HH}\alpha \geq \begin{cases} \delta_{HL}\delta\alpha, & \text{if } p_{HL}=1 \text{ and } n_{HL} \text{ even} \\ \delta_{HL}\delta^2\alpha, & \text{if } p_{HL}=1 \text{ and } n_{HL} \text{ odd} \\ \overset{\bullet}{\delta}_{HL}\delta\alpha, & \text{if } p_{HL}=0 \end{cases} \quad \text{and } F(H) \text{ and } t_{HL}<t_{HH}$$

$$\tag{CR1}$$

$$p_{HL}\delta_{HL}a_{HL} + \overset{\bullet}{p}_{HL}\overset{\bullet}{\delta}_{HL}\alpha \geq \begin{cases} \delta_{HH}\delta\alpha, & \text{if } p_{HH}=1 \text{ and } n_{HH} \text{ even} \\ \delta_{HH}\delta^2\alpha, & \text{if } p_{HH}=1 \text{ and } n_{HH} \text{ odd} \\ \overset{\bullet}{\delta}_{HH}\delta\alpha, & \text{if } p_{HH}=0 \end{cases} \quad \text{and } F(H) \text{ and } t_{HL}>t_{HH}$$

$$\tag{CR2}$$

$$p_{HH}\delta_{HH}(100-a_{HH})+p_{HH}^{*}\delta_{HH}^{*}\alpha \geq \begin{cases} \delta_{LH}\delta\alpha, & \text{if } p_{LH}=1 \text{ and } n_{LH} \text{ odd} \\ \delta_{LH}\delta^2\alpha, & \text{if } p_{LH}=1 \text{ and } n_{LH} \text{ even} \\ \delta_{LH}^{*}\delta\alpha, & \text{if } p_{LH}=0 \end{cases} \text{and } S(H) \text{ and } t_{LH}<t_{HH}$$

(CR3)

$$p_{LH}\delta_{LH}(100-a_{LH})+p_{LH}^{*}\delta_{LH}^{*}\alpha \geq \begin{cases} \delta_{HH}\delta\alpha, & \text{if } p_{HH}=1 \text{ and } n_{HH} \text{ odd} \\ \delta_{HH}\delta^2\alpha, & \text{if } p_{HH}=1 \text{ and } n_{HH} \text{ even} \\ \delta_{HH}^{*}\delta\alpha, & \text{if } p_{HH}=0 \end{cases} \text{and } S(H) \text{ and } t_{LH}>t_{HH}$$

(CR4)

Suppose $F(H)$ and $t_{HL}<t_{HH}$. This means type H_1 can distinguish between both types of the opponent in step t_{HL}. Further suppose $p_{HL}=1$ and n_{HL} even, this means H_1 and L_2 agree and the agreement allocation is proposed by L_2. If L_2 did not propose the agreement allocation type H_1 knows that his opponent is strong too. Then the continuation of the play with the strong type should be at least as profitable as a break off by type H_1 (in decision step $n_{HL}+1$). If $p_{HL}=1$ and n_{HL} odd, this means that type H_1 is supposed to propose the final agreement with L_2, a unilateral break off of H_1 is possible in step $n_{HL}+2$, if his proposal was not accepted and he is therefore aware of a strong opponent. If $p_{HL}^{*}=1$, this means that H_1 and L_2 do not agree, a continuation constraint for H_1 can only be deduced if n_{HL}^{*} even, this means that it is not H_1 who initiates the break off. The continuation individual rationality constraint (CR1) expresses that in all three cases a continuation of the game is as least as profitable as a unilateral break off by H_1. Conditions (CR2) covers the case that type H_1 is aware of a weak opponent and the other two constraints state these conditions for the strong second mover.

CONTINUATION INCENTIVE CONSTRAINTS

If a player gained information about the opponent's type (this means F(i) or S(j) is fulfilled), it should be more profitable for him to continue playing according to his own strategy than imitating the other type in a play with this particular opponent. The incentive constraints excluded a complete imitation of the other type, the *continuation incentive constraints* exclude the partial imitation in the continuation of the game after a player gained information about the opponent's type. This imitation of the other type would only be partial concerning that part of the other type's strategy in a play with the particular opponent. A strategy not satisfying the continuation incentive constraints cannot be an equilibrium strategy, since it offers the players an incentive to deviate during the play.

Suppose S(H). This means that the high type second mover is able to distinguish between the two types of the opponent. Furthermore suppose $t_{LH} < t_{HH}$, which means that the termination of a play with L_1 occurs before the termination of the play with H_1. In case the game does not end by acceptance or break off in step t_{LH}, the player knows that he faces a strong opponent. Then, for the continuation of the game a play according to the strategy of H_2 has to be at least as profitable as an imitation of the part of L_2 in a play with H_1.

$$p_{HH}\delta_{HH}(100-a_{HH}) + \overset{\bullet}{p}_{HH}\overset{\bullet}{\delta}_{HH}\alpha \geq p_{HL}\delta_{HL}(100-a_{HL}) + \overset{\bullet}{p}_{HL}\overset{\bullet}{\delta}_{HL}\alpha, \text{ if } S(H) \text{ and } t_{LH}<t_{HH} \quad (CI1)$$

$$p_{LH}\delta_{LH}(100-a_{LH}) + \overset{\bullet}{p}_{LH}\overset{\bullet}{\delta}_{LH}\alpha \geq p_{LL}\delta_{LL}(100-a_{LL}) + \overset{\bullet}{p}_{LL}\overset{\bullet}{\delta}_{LL}\alpha, \text{ if } S(H) \text{ and } t_{LH}>t_{HH} \quad (CI2)$$

$$p_{HL}\delta_{HL}(100-a_{HL}) \geq p_{HH}\delta_{HH}(100-a_{HH}), \text{ if } S(L) \text{ and } t_{LL}<t_{HL} \quad (CI3)$$

$$p_{LL}\delta_{LL}(100-a_{LL}) \geq p_{LH}\delta_{LH}(100-a_{LH}), \text{ if } S(L) \text{ and } t_{LL}>t_{HL} \quad (CI4)$$

$$p_{HH}\delta_{HH}a_{HH} + \overset{\bullet}{p}_{HH}\overset{\bullet}{\delta}_{HH}\alpha \geq p_{LH}\delta_{LH}a_{LH} + \overset{\bullet}{p}_{LH}\overset{\bullet}{\delta}_{LH}\alpha, \text{ if } F(H) \text{ and } t_{HL}<t_{HH} \quad (CI5)$$

$$p_{HL}\delta_{HL}a_{HL} + \overset{\bullet}{p}_{HL}\overset{\bullet}{\delta}_{HL}\alpha \geq p_{LL}\delta_{LL}a_{LL} + \overset{\bullet}{p}_{LL}\overset{\bullet}{\delta}_{LL}\alpha, \text{ if } F(H) \text{ and } t_{HL}>t_{HH} \quad (CI6)$$

$$p_{LH}\delta_{LH}a_{LH} \geq p_{HH}\delta_{HH}a_{HH}, \text{ if } F(L) \text{ and } t_{LL}<t_{LH} \quad (CI7)$$

$$p_{LL}\delta_{LL}a_{LL} \geq p_{HL}\delta_{HL}a_{HL}, \text{ if } F(L) \text{ and } t_{LL}>t_{LH} \quad (CI8)$$

The following smallest money unit constraints are also partially a continuation constraints.

Smallest Money Unit Constraints

Suppose that a low type first mover has an agreement with both types of the second mover and that he can distinguish between the two types of the opponent, this means F(L) is satisfied. Suppose further $n_{LL} < n_{LH}$ and n_{LL} even, this means that the agreement with the low type opponent is reached before the agreement with the high type opponent and that it is proposed by the low type second mover. If the opponent does not demand $100 - a_{LL}$ in n_{LL} the weak type first mover knows that his opponent is strong. In this case the *smallest money unit constraints* demand that the agreement payoff $\delta_{LH} a_{LH}$ in a play with this opponent has to be at least δ_{LL}. If this would not be the case, L_1 would have the incentive to accept every proposal by H_2 in step n_{LL}. The demands have to be multiples of the smallest money unit 1, this means they have to be integer values. Therefore by acceptance of the proposal in step n_{LL} type L_1 receives at least $\delta_{LL} \cdot 1$. The continuation according to the prescribed strategy has to be at least as profitable as the acceptance in the step of the distinction of the two types of the opponent.

If a weak player only agrees with one type of the opponent, this payoff has to be at least 1, since otherwise the weak player has an incentive to accept every first round proposal (in case he has the opportunity to do this).

A set of strategies not satisfying the smallest money unit constraints cannot be in equilibrium, since the low type would have an incentive to deviate.

The smallest money unit constraints (M1) for the low type first mover and (M2) for the low type second mover are as follows.

$$\left.\begin{array}{l} \delta_{LL}a_{LL} \geq 1, \text{ if } p_{LL}=1 \text{ and } p_{LH}=0 \\[4pt] \delta_{LH}a_{LH} \geq 1, \text{ if } p_{LL}=0 \text{ and } p_{iH}=1 \\[4pt] \delta_{LL}a_{LL} \geq 1, \text{ if } p_{LL}=1 \text{ and } p_{LH}=1 \text{ and } n_{LL}=n_{LH} \\[4pt] \delta_{LH}a_{LH} \geq 1, \text{ if } p_{LL}=1 \text{ and } p_{LH}=1 \text{ and } n_{LL}=n_{LH} \\[4pt] \delta_{LH}a_{LH} \geq \delta_{LL}, \text{ if } p_{LL}=1 \text{ and } p_{LH}=1 \text{ and } n_{LL}<n_{LH} \text{ and } n_{LL} \text{ even} \\[4pt] \delta_{LH}a_{LH} \geq \delta\delta_{LL}, \text{ if } p_{LL}=1 \text{ and } p_{LH}=1 \text{ and } n_{LL}<n_{LH} \text{ and } n_{LL} \text{ odd} \\[4pt] \delta_{LL}a_{LL} \geq \delta_{LH}, \text{ if } p_{LL}=1 \text{ and } p_{LH}=1 \text{ and } n_{LL}>n_{LH} \text{ and } n_{LH} \text{ even} \\[4pt] \delta_{LL}a_{LL} \geq \delta\delta_{LH}, \text{ if } p_{LL}=1 \text{ and } p_{LH}=1 \text{ and } n_{LL}>n_{LH} \text{ and } n_{LH} \text{ odd} \end{array}\right\} \quad \text{(M1)}$$

$$\left.\begin{array}{l} \delta_{LL}(100-a_{LL}) \geq 1, \text{ if } p_{LL}=1 \text{ and } p_{HL}=0 \\[4pt] \delta_{HL}(100-a_{HL}) \geq 1, \text{ if } p_{LL}=0 \text{ and } p_{HL}=1 \\[4pt] \delta_{LL}(100-a_{LL}) \geq 1, \text{ if } p_{LL}=1 \text{ and } p_{HL}=1 \text{ and } n_{LL}=n_{HL} \\[4pt] \delta_{HL}(100-a_{HL}) \geq 1, \text{ if } p_{LL}=1 \text{ and } p_{HL}=1 \text{ and } n_{LL}=n_{HL} \\[4pt] \delta_{HL}(100-a_{HL}) \geq \delta_{LL}, \text{ if } p_{LL}=1 \text{ and } p_{HL}=1 \text{ and } n_{LL}<n_{HL} \text{ and } n_{LL} \text{ odd} \\[4pt] \delta_{HL}(100-a_{HL}) \geq \delta\delta_{LL}, \text{ if } p_{LL}=1 \text{ and } p_{HL}=1 \text{ and } n_{LL}<n_{HL} \text{ and } n_{LL} \text{ even} \\[4pt] \delta_{LL}(100-a_{LL}) \geq \delta_{HL}, \text{ if } p_{LL}=1 \text{ and } p_{HL}=1 \text{ and } n_{LL}>n_{HL} \text{ and } n_{HL} \text{ odd} \\[4pt] \delta_{LL}(100-a_{LL}) \geq \delta\delta_{HL}, \text{ if } p_{LL}=1 \text{ and } p_{HL}=1 \text{ and } n_{LL}>n_{HL} \text{ and } n_{HL} \text{ even} \end{array}\right\} \quad \text{(M2)}$$

DELAY CONSTRAINTS

Two *delay constraints* emerge from the fact that a weak player is better off by accepting any arbitrary proposal than by having a conflict. Hence, a strategy which prescribes a conflict with a low type after the low type received an offer by the opponent cannot be an equilibrium strategy, since the low type has the incentive to deviate from the break off prescription and to accept the proposal.

If $p_{LL}=0$ then $\delta_{LL}^{*}=1$.   (D1)

If $p=(0,\cdot,0,1)$ then $\delta_{LH}^{*}=\delta$.   (D2)

If $p=(0,1,\cdot,\cdot)$ then $\delta_{HH}^{*}\leq\delta$.   (D3)

The " $\cdot$ " stands for an arbitrary value from $\{0,1\}$.

The first constraint expresses that whenever a strategy prescribes that two weak types have a conflict, this has to be immediately. Otherwise, the strategy cannot be an equilibrium strategy, since a low type would have an incentive to deviate and to accept a proposal.

The second constraint says that if H_2 is going to have a conflict with each of both types of the opponent, but two low type players are going to agree, the high type second mover has to break off immediately. Since the two low types are prescribed to have an agreement the low type first mover cannot break off in his first decision step. A demand of H_2 before his break off would be an incentive for L_1 to accept since this is more profitable than a conflict with H_2. Therefore, a set of strategies for this vector p which allows H_2 to demand before the break off cannot be in equilibrium.

Constraint (D3) is a feasibility constraint. If the high type first mover is having a conflict with the high type second mover and an agreement with the low type second mover he cannot break off immediately.

SET OF THE NECESSARY CONSTRAINTS

With the term *set of the necessary constraints* we shall shortly denote the individual rationality constraints (R1) and (R2), the incentive constraints (I1) to (I4), the partial individual rationality constraints (PR1) and (PR2), the continuation individual rationality constraints (CR1) to (CR4), the continuation incentive constraints (PI1) to (PI8), the smallest money unit constraints (M1) and (M2), and the delay constraints (D1) to (D3).

Theorem 1:

(I)   Every parameter constellation belonging to an equilibrium point in pure and finite strategies of the two-person bargaining game with incomplete information has to fulfill the set of the necessary constraints.

(II)   For every parameter constellation of a_{ij}, p_{ij}, and $p_{ij}n_{ij} + p_{ij}^{*}n_{ij}^{*}$ in the admissible parameter range satisfying the set of the necessary constraints an equilibrium in pure and finite strategies can be constructed, $i,j \in \{H,L\}$.

Proof:

(I)   With the presentation of the constraints it was argued why they are necessary for the parameter constellations associated with equilibrium strategies. They put restrictions on the parameters such that ex ante, as well as in the course of the bargaining it is not profitable for a player to deviate from his strategy and to terminate the play unilaterally, or imitate the opponent. The delay constraint is necessary for equilibrium strategies, since the alternative of 0 makes it profitable for the low type to accept any arbitrary proposal instead of having a conflict. The smallest money unit constraints are necessary since the demands have to be integers and therefore the agreement payoffs of the weak types cannot be too small.

(II)   We shall proof the claim by the distinction of 16 possible vectors of p, which emerge since the p_{ij} can only take the values 0 and 1.

The set of the necessary constraints is not fulfilled for ten of the combinations of agreement probabilities. There exist restrictions which are in contradiction with either the admissible ranges of the variables or with other conditions. Table 3.1 shows eight of the combinations of agreement probabilities for which the contradiction emerges from the incentive constraints.

Table 3.1: Combinations of the agreement probabilities which do not fulfill the set of the necessary constraints

p	Restrictions in contradiction
(0,0,1,0)	(I4): $a_{LH} \geq 100$
(0,1,0,0)	(I3): $a_{HL} \leq 0$
(1,0,0,0)	(I3)&(I4): $100 \leq a_{HH} \leq 0$
(1,0,1,0)	(I4): $\delta_{HH}(100-a_{HH})+\delta_{LH}(100-a_{LH}) \leq 0$
(1,1,0,0)	(I3): $\delta_{HH}a_{HH}+\delta_{HL}a_{HL} \leq 0$
(1,0,1,1)	(I2)&(I4): $\delta_{LL}(100-a_{LL})+\delta_{HL}^{*}\alpha \leq \delta_{HH}(100-a_{HH})+\delta_{LH}(100-a_{LH}) \leq \delta_{LL}(100-a_{LL})$
(1,1,0,1)	(I1)&(I3): $\delta_{LL}a_{LL}+\delta_{LH}^{*}\alpha \leq \delta_{HH}a_{HH}+\delta_{HL}a_{HL} \leq \delta_{LL}a_{LL}$
(1,1,1,0)	(I1)&(I3): $\delta_{LH}a_{LH}+\delta_{LL}^{*}\alpha \leq \delta_{HH}a_{HH}+\delta_{HL}a_{HL} \leq \delta_{LH}a_{LH}$

For two further combinations of the agreement probabilities it is more complicated to show that the set of the necessary constraints is not satisfied.

Case p=(1,0,0,1):

The individual rationality and the incentive constraints yield:

$$\delta_{HH}a_{HH} + \delta_{HL}^{*}\alpha \geq 2\alpha \tag{1}$$

$$\delta_{HH}(100-a_{HH}) + \delta_{LH}^{*}\alpha \geq 2\delta\alpha \tag{2}$$

$$\delta_{HH}a_{HH} + \delta_{HL}^{*}\alpha \geq \delta_{LL}a_{LL} + \delta_{LH}^{*}\alpha \tag{3}$$

$$\delta_{LL}a_{LL} \geq \delta_{HH}a_{HH} \tag{4}$$

$$\delta_{HH}(100-a_{HH}) + \delta_{LH}^{*}\alpha \geq \delta_{LL}(100-a_{LL}) + \delta_{HL}^{*}\alpha \tag{5}$$

$$\delta_{LL}(100-a_{LL}) \geq \delta_{HH}(100-a_{HH}) \tag{6}$$

Inequalities (3) and (5) yield

$$\delta_{HH} \geq \delta_{LL}. \tag{7}$$

Inequalities (4) and (6) yield

$$\delta_{HH} \leq \delta_{LL}. \tag{8}$$

Therefore it follows

$$\delta_{HH} = \delta_{LL}. \tag{9}$$

Inequalities (4) and (6) together with (9) yield

$$a_{LL} = a_{HH}. \tag{10}$$

And inequalities (3) and (5) together with (9) and (10) yield

$$\delta_{HL}^{\bullet} = \delta_{LH}^{\bullet}. \tag{11}$$

With help of equality (11) the inequalities (1) and (2) yield

$$\delta_{LH}^{\bullet} = (1+\delta)\frac{50}{\alpha}\delta_{LL}. \tag{12}$$

Together with (9) and (10) the partial individual rationality constraints (PR1) and (PR2) yield

$$\frac{50}{\alpha} \geq 1+\delta. \tag{13}$$

Therefore (12) yields

$$\delta_{LH}^{\bullet} \geq \delta_{LL}, \tag{14}$$

which is equivalent to

$$t_{LL} < t_{LH}. \tag{15}$$

Since F(L) and S(L) are satisfied, constraints (CI7) and (CI3) have to be satisfied:

(CI7): $0 \geq \delta_{HH}a_{HH}.$ (16)

(CI3): $0 \geq \delta_{HH}(100-a_{HH}).$ (17)

But, inequalities (16) and (17) are in contradiction to the admissible parameter range, since both components of the product are strictly positive. This means that for p=(1,0,0,1) the set of the necessary constraints cannot be fulfilled.

Case (0,1,1,0):

The individual rationality and the incentive constraints yield:

$$\delta_{HL}a_{HL} + \delta_{HH}^{\bullet}\alpha \geq 2\alpha \tag{18}$$

$$\delta_{LH}(100-a_{LH}) + \delta_{HH}^{\bullet}\alpha \geq 2\delta\alpha \tag{19}$$

$$\delta_{HL}a_{HL} + \delta_{HH}^{\bullet}\alpha \geq \delta_{LH}a_{LH} + \delta_{LL}^{\bullet}\alpha \tag{20}$$

$$\delta_{LH}a_{LH} \geq \delta_{HL}a_{HL} \tag{21}$$

$$\delta_{LH}(100-a_{LH}) + \delta_{HH}^{*}\alpha \geq \delta_{HL}(100-a_{HL}) + \delta_{LL}^{*}\alpha \tag{22}$$

$$\delta_{HL}(100-a_{HL}) \geq \delta_{LH}(100-a_{LH}) \tag{23}$$

Constraint (D1) yields

$$\delta_{LL}^{*} = 1, \tag{24}$$

and constraint (D3) yields

$$\delta_{HH}^{*} \leq \delta. \tag{25}$$

From inequalities (20) and (21) it follows

$$\delta_{HH}^{*} \geq \delta_{LL}^{*}. \tag{26}$$

Together with (24) this implies

$$\delta_{HH}^{*} = 1. \tag{27}$$

But this is in contradiction to inequality (25). This means that for $p=(0,1,1,0)$ the set of the necessary constraints is in contradiction.

For the other six parameter constellations of p (shown in table 3.2) pure and finite equilibrium strategies can be constructed.

Table 3.2: Combinations of agreement probabilities which satisfy the set of the necessary constraints

Case	p_{HH}	p_{HL}	p_{LH}	p_{LL}
1	0	0	0	0
2	0	0	0	1
3	0	0	1	1
4	0	1	0	1
5	0	1	1	1
6	1	1	1	1

Let us first introduce some conventions used in the description of the strategies. A *decision step* of a player is a step where it is the player's turn to decide. The decision steps of the first mover are all odd steps, while the even steps are the decision steps of the second mover. The request "*demand x before step n*" means that the player should demand the value x in all his decision steps 1 to $n-1$. The request "*demand x between step n and step m*" expresses that the player should demand x in all decision steps $n+1$ to $m-1$. The request "*demand x after step m*" means that the player should demand x in all decision steps from $m+1$ on. The requests are executed only if the number of steps prescribed is at least 1. All three requests are executed regardless of previous history.

We shall construct the strategies in such a way that delay is caused by demands of 99. By the necessary constraints (M1) and (M2) it is ensured that a low type player does not have an incentive to deviate and accept the 99. The partial individual rationality constraints ensure that the high type is not going to accept the 99.

<u>Case 1:</u> $p_{HH} = 0$, $p_{HL} = 0$, $p_{LH} = 0$, $p_{LL} = 0$
The set of the necessary constraints yields:

$$\delta^*_{HH} = 1 \qquad\qquad\qquad\qquad\qquad\qquad\qquad\qquad\qquad (1.1)$$

$$\delta^*_{HL} = 1 \qquad\qquad\qquad\qquad\qquad\qquad\qquad\qquad\qquad (1.2)$$

$$\delta^*_{LH} = 1 \qquad\qquad\qquad\qquad\qquad\qquad\qquad\qquad\qquad (1.3)$$

$$\delta^*_{LL} = 1 \qquad\qquad\qquad\qquad\qquad\qquad\qquad\qquad\qquad (1.4)$$

This means that all four plays have to end immediately by a break off.

<u>A set of finite and pure equilibrium strategies</u>
H_1:   *Break off in each decision step*
L_1:   *Break off in each decision step*
H_2:   *Break off in each decision step*
L_2:   *Break off in each decision step*

These strategies are equilibrium strategies which are uniquely determined by the necessary conditions. For no type a unilateral deviation is profitable. This set of equilibrium strategies is implementable for all parameter constellations in the admissible range, satisfying the necessary conditions.

The expected payoff of a type does not depend on his position in the mover sequence.

$$P_H^1 = \alpha \tag{1.5}$$

$$P_H^2 = \alpha \tag{1.6}$$

$$P_L^1 = 0 \tag{1.7}$$

$$P_L^2 = 0 \tag{1.8}$$

Case 2: $p_{HH} = 0$, $p_{HL} = 0$, $p_{LH} = 0$, $p_{LL} = 1$
The set of the necessary constraints yields:

$$\delta_{HH}^* = 1 \tag{2.1}$$

$$\delta_{HL}^* = 1 \tag{2.2}$$

$$\delta_{LH}^* = \delta \tag{2.3}$$

$$1 \leq \delta_{LL} a_{LL} \leq (2-\delta)\alpha \tag{2.4}$$

$$1 \leq \delta_{LL}(100-a_{LL}) \leq \delta\alpha \tag{2.5}$$

Inequalities (2.4) and (2.5) imply

$$\alpha \geq 50\delta_{LL}. \tag{2.6}$$

A set of finite and pure equilibrium strategies (for n_{LL} odd)
H_1: *Break off in each decision step*
L_1: *Demand and accept 99 before step n_{LL}, demand and accept a_{LL} in step n_{LL}, break off after step n_{LL}*
H_2: *Break off in each decision step*
L_2: *Demand and accept 99 before step $n_{LL}+1$, demand and accept $100-a_{LL}$ in step $n_{LL}+1$,*

This set of equilibrium strategies is constructed for n_{LL} odd. This means that the first mover proposes the final agreement. An analogous set of equilibrium strategies can be constructed for n_{LL} even, where the final agreement is proposed by the second mover. Then it has to be n_{LL} (instead of $n_{LL}+1$) in the strategy of L_2 and $n_{LL}+1$ (instead of n_{LL}) in the strategy of L_1.

For all admissible parameter constellations which satisfy the set of the necessary constraints this set of strategies is in equilibrium.

The expected payoffs are

$$P_H^1 = \alpha \tag{2.7}$$

$$P_H^2 = .5 \cdot (1+\delta)\alpha \tag{2.8}$$

$$P_L^1 = .5 \cdot \delta_{LL} a_{LL} \tag{2.9}$$

$$P_L^2 = .5 \cdot \delta_{LL}(100 - a_{LL}) \tag{2.10}$$

Case 3: $p_{HH} = 0$, $p_{HL} = 0$, $p_{LH} = 1$, $p_{LL} = 1$

The set of the necessary constraints yields:

$$\delta_{HH}^* = 1 \tag{3.1}$$

$$\delta_{HL}^* = 1 \tag{3.2}$$

$$\delta_{LL}(100 - a_{LL}) = \delta_{LH}(100 - a_{LH}) \tag{3.3}$$

$$\delta_{LH}(100 - a_{LH}) \geq \delta\alpha \tag{3.4}$$

$$\delta_{LL}a_{LL} + \delta_{LH}a_{LH} \leq 2\alpha \tag{3.5}$$

$$100 - a_{LH} \geq \alpha, \text{ if } n_{LH} \text{ even} \tag{3.6}$$

$$100 - a_{LH} \geq \delta\alpha, \text{ if } n_{LH} \text{ odd} \tag{3.7}$$

$$\delta_{LL}a_{LL} \geq 1, \text{ if } n_{LL} = n_{LH} \tag{3.8}$$

$$\delta_{LH}a_{LH} \geq 1, \text{ if } n_{LL} = n_{LH} \tag{3.9}$$

$$\delta_{LH} a_{LH} \geq \delta_{LL}, \text{ if } n_{LL} < n_{LH} \text{ and } n_{LL} \text{ even} \tag{3.10}$$

$$\delta_{LH} a_{LH} \geq \delta \delta_{LL}, \text{ if } n_{LL} < n_{LH} \text{ and } n_{LL} \text{ odd} \tag{3.11}$$

$$\delta_{LL} a_{LL} \geq \delta_{LH}, \text{ if } n_{LL} > n_{LH} \text{ and } n_{LH} \text{ even} \tag{3.12}$$

$$\delta_{LL} a_{LL} \geq \delta \delta_{LH}, \text{ if } n_{LL} > n_{LH} \text{ and } n_{LH} \text{ odd} \tag{3.13}$$

Inequality (3.3) expresses that it is necessary, that both types of the second mover receive the same (discounted) payoff, although the outcomes $100 - a_{LL}$ and $100 - a_{LH}$ might be different.

A set of finite and pure equilibrium strategies (for n_{LL} odd, n_{LH} odd, and $n_{LL} \leq n_{LH}$)

H_1:   *Break off in each step*

L_1:   *Demand and accept 99 before step n_{LL}, demand and accept a_{LL} in step n_{LL}, demand and*
     accept 99 in the steps between n_{LL} and n_{LH}, demand and accept a_{LH} in step n_{LH}, break
     off after step n_{LH}

H_2:   *Demand and accept 99 before step $n_{LH}+1$, demand and accept $100 - a_{LH}$ in step $n_{LH}+1$,*
     break off after step $n_{LH}+1$

L_2:   *Demand and accept 99 before step $n_{LL}+1$, demand and accept $100 - a_{LL}$ in step $n_{LL}+1$,*
     break off after step $n_{LL}+1$

This set of equilibrium strategies is constructed for n_{LL} odd and n_{LH} odd. This means that the first mover proposes the final agreement in both cases. Analogously, like discussed in case 2, sets of equilibrium strategies can be constructed for all four possible combinations of n_{LL} and n_{LH} even and odd, respectively.

Furthermore, the strategies are constructed such that the two low type players agree before the low type first mover and the high type second mover on the equilibrium path. Obviously, a set of equilibrium strategies can also be constructed for $n_{LL} \geq n_{LH}$. Then the two variables (together with the respective demands) simply have to be exchanged in the strategy prescription of L_1.

For all admissible parameter constellations which satisfy the set of the necessary constraints this set of strategies is in equilibrium.

The expected payoffs are

$$P_H^1 = \alpha \tag{3.14}$$

$$P_H^2 = .5 \cdot (\delta_{LH}(100-a_{LH}) + \alpha) \tag{3.15}$$

$$P_L^1 = .5 \cdot (\delta_{LH}a_{LH} + \delta_{LL}a_{LL}) \tag{3.16}$$

$$P_L^2 = .5 \cdot \delta_{LL}(100-a_{LL}) \tag{3.17}$$

<u>Case 4:</u> $p_{HH} = 0$, $p_{HL} = 1$, $p_{LH} = 0$, $p_{LL} = 1$

The set of the necessary constraints yields:

$$\delta_{HH}^* = \delta \tag{4.1}$$

$$\delta_{LH}^* = \delta \tag{4.2}$$

$$\delta_{LL}a_{LL} = \delta_{HL}a_{HL} \tag{4.3}$$

$$\delta_{HL}a_{HL} \geq (2-\delta)\alpha \tag{4.4}$$

$$\delta_{HL}(100-a_{HL}) + \delta_{LL}(100-a_{LL}) \leq 2\delta\alpha \tag{4.5}$$

$$a_{HL} \geq \alpha, \text{ if } n_{HL} \text{ odd} \tag{4.6}$$

$$a_{HL} \geq \delta\alpha, \text{ if } n_{HL} \text{ even} \tag{4.7}$$

$$\delta_{LL}(100-a_{LL}) \geq 1, \text{ if } n_{LL}=n_{HL} \tag{4.8}$$

$$\delta_{HL}(100-a_{HL}) \geq 1, \text{ if } n_{LL}=n_{HL} \tag{4.9}$$

$$\delta_{HL}(100-a_{HL}) \geq \delta_{LL}, \text{ if } n_{LL}<n_{HL} \text{ and } n_{LL} \text{ odd} \tag{4.10}$$

$$\delta_{HL}(100-a_{HL}) \geq \delta\delta_{LL}, \text{ if } n_{LL}<n_{HL} \text{ and } n_{LL} \text{ even} \tag{4.11}$$

$$\delta_{LL}(100-a_{LL}) \geq \delta_{HL}, \text{ if } n_{LL}>n_{HL} \text{ and } n_{HL} \text{ odd} \tag{4.12}$$

$$\delta_{LL}(100-a_{LL}) \geq \delta\delta_{HL}, \text{ if } n_{LL}>n_{HL} \text{ and } n_{HL} \text{ even} \tag{4.13}$$

From inequality (4.3) it follows that both first mover types receive the same (discounted) payoff in a play with the low type of the second mover, although the outcomes a_{LL} and a_{HL} might be different.

<u>A set of finite and pure equilibrium strategies</u> (for n_{HL} odd, n_{LL} odd, $n_{LL} \leq n_{HL}$)

H_1: *Demand and accept 99 before step n_{HL}, demand and accept a_{HL} in step n_{HL}, break off after step n_{HL}*

L_1: *Demand and accept 99 before step n_{LL}, demand and accept a_{LL} in step n_{LL}, break off after step n_{LL}*

H_2: *Break off in each step*

L_2: *Demand and accept 99 before step $n_{LL}+1$, demand and accept $100-a_{LL}$ in step $n_{LL}+1$, demand and accept 99 in the steps between $n_{LL}+1$ and $n_{HL}+1$, demand and accept $100-a_{HL}$ in step $n_{HL}+1$, break off after step $n_{HL}+1$*

This set of equilibrium strategies is constructed for n_{LL} odd and n_{HL} odd. This means that the first mover proposes the final agreement in both cases. Analogously, like discussed in case 2, sets of equilibrium strategies can be constructed for all four possible combinations of n_{LL} and n_{HL} even and odd, respectively.

Furthermore, the strategies are constructed such that the two low type players agree before the high type first mover and the low type second mover on the equilibrium path. Obviously, a set of equilibrium strategies can also be constructed for $n_{LL} \geq n_{HL}$ (as discussed in case 3).

For all admissible parameter constellations which satisfy the set of the necessary constraints this set of strategies is in equilibrium.

The expected payoffs are

$$P_H^1 = .5 \cdot (\delta_{HL} a_{HL} + \delta\alpha) \tag{4.14}$$

$$P_H^2 = \delta\alpha \tag{4.15}$$

$$P_L^1 = .5 \cdot \delta_{LL} a_{LL} \tag{4.16}$$

$$P_L^2 = .5 \cdot (\delta_{HL}(100-a_{HL}) + \delta_{LL}(100-a_{LL})) \tag{4.17}$$

<u>Case 5:</u> $p_{HH} = 0$, $p_{HL} = 1$, $p_{LH} = 1$, $p_{LL} = 1$

The set of the necessary constraints yields:

$$\delta_{HL}a_{HL} \geq (2-\delta_{HH}^{*})\alpha \tag{5.1}$$

$$\delta_{LH}(100-a_{LH}) \geq (2\delta-\delta_{HH}^{*})\alpha \tag{5.2}$$

$$\delta_{HL}a_{HL} \leq \delta_{LH}a_{LH} + \delta_{LL}a_{LL} \leq \delta_{HL}a_{HL} + \delta_{HH}^{*}\alpha \tag{5.3}$$

$$\delta_{LH}(100-a_{LH}) \leq \delta_{HL}(100-a_{HL}) + \delta_{LL}(100-a_{LL}) \leq \delta_{LH}(100-a_{LH}) + \delta_{HH}^{*}\alpha \tag{5.4}$$

$$\delta_{LH}a_{LH} \leq \delta_{HH}^{*}\alpha, \text{ if } n_{HL}<n_{HH}^{*} \tag{5.5}$$

$$\delta_{HL}a_{HL} \geq \delta_{LL}a_{LL}, \text{ if } n_{HL}>n_{HH}^{*} \text{ and } n_{HH}^{*} \text{ even} \tag{5.6}$$

$$\delta_{LH}(100-a_{LH}) \geq \delta_{LL}(100-a_{LL}), \text{ if } n_{LH}>n_{HH}^{*} \text{ and } n_{HH}^{*} \text{ odd} \tag{5.7}$$

$$\delta_{HL}(100-a_{HL}) \leq \delta_{HH}^{*}\alpha, \text{ if } n_{LH}<n_{HH}^{*} \tag{5.8}$$

$$\delta_{LL}a_{LL} \geq \delta_{HL}a_{HL}, \text{ if } n_{LL}>n_{LH} \tag{5.9}$$

$$\delta_{LL}(100-a_{LL}) \geq \delta_{LH}(100-a_{LH}), \text{ if } n_{LL}>n_{HL} \tag{5.10}$$

$$\delta_{HH}^{*} \geq \delta_{HL}\delta, \text{ if } n_{HH}^{*}>n_{HL} \text{ and } n_{HL} \text{ even} \tag{5.11}$$

$$\delta_{HH}^{*} \geq \delta_{HL}\delta^{2}, \text{ if } n_{HH}^{*}>n_{HL} \text{ and } n_{HL} \text{ odd} \tag{5.12}$$

$$\delta_{HL}a_{HL} \geq \delta_{HH}^{*}\delta\alpha, \text{ if } n_{HH}^{*}<n_{HL} \text{ and } n_{HH}^{*} \text{ even} \tag{5.13}$$

$$\delta_{HH}^{*} \geq \delta_{LH}\delta, \text{ if } n_{HH}^{*}>n_{LH} \text{ and } n_{LH} \text{ odd} \tag{5.14}$$

$$\delta_{HH}^{*} \geq \delta_{LH}\delta^{2}, \text{ if } n_{HH}^{*}>n_{LH} \text{ and } n_{LH} \text{ even} \tag{5.15}$$

$$\delta_{LH}(100-a_{LH}) \geq \delta_{HH}^{*}\delta\alpha, \text{ if } n_{HH}^{*}<n_{LH} \text{ and } n_{HH}^{*} \text{ odd} \tag{5.16}$$

$$a_{HL} \geq \alpha, \text{ if } n_{HL} \text{odd} \tag{5.17}$$

$$a_{HL} \geq \delta\alpha, \text{ if } n_{HL} \text{even} \tag{5.18}$$

$$100-a_{LH} \geq \alpha, \text{ if } n_{LH} \text{even} \tag{5.19}$$

$$100-a_{LH} \geq \delta\alpha, \text{ if } n_{LH} \text{odd} \tag{5.20}$$

$$\delta_{LL}a_{LL} \geq 1, \text{ if } n_{LL}=n_{LH} \tag{5.21}$$

$$\delta_{LH}a_{LH} \geq 1, \text{ if } n_{LL}=n_{LH} \tag{5.22}$$

$$\delta_{LH}a_{LH} \geq \delta_{LL}, \text{ if } n_{LL}<n_{LH} \text{ and } n_{LL} \text{ even} \tag{5.23}$$

$$\delta_{LH}a_{LH} \geq \delta\delta_{LL}, \text{ if } n_{LL}<n_{LH} \text{ and } n_{LL} \text{ odd} \tag{5.24}$$

$\delta_{LL}a_{LL} \geq \delta_{LH}$, if $n_{LL}>n_{LH}$ and n_{LH} even $\hfill$ (5.25)

$\delta_{LL}a_{LL} \geq \delta\delta_{LH}$, if $n_{LL}>n_{LH}$ and n_{LH} odd $\hfill$ (5.26)

$\delta_{LL}(100-a_{LL}) \geq 1$, if $n_{LL}=n_{HL}$ $\hfill$ (5.27)

$\delta_{HL}(100-a_{HL}) \geq 1$, if $n_{LL}=n_{HL}$ $\hfill$ (5.28)

$\delta_{HL}(100-a_{HL}) \geq \delta_{LL}$, if $n_{LL}<n_{HL}$ and n_{LL} odd $\hfill$ (5.29)

$\delta_{HL}(100-a_{HL}) \geq \delta\delta_{LL}$, if $n_{LL}<n_{HL}$ and n_{LL} even $\hfill$ (5.30)

$\delta_{LL}(100-a_{LL}) \geq \delta_{HL}$, if $n_{LL}>n_{HL}$ and n_{HL} odd $\hfill$ (5.31)

$\delta_{LL}(100-a_{LL}) \geq \delta\delta_{HL}$, if $n_{LL}>n_{HL}$ and n_{HL} even $\hfill$ (5.32)

The following diagram shall illustrate an equilibrium path in a highly stylized way. Only the demands that yield the agreements are shown.

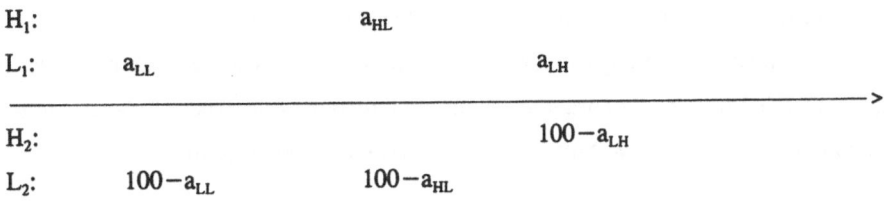

H_1: a_{HL}

L_1: a_{LL} a_{LH}

H_2: $100-a_{LH}$

L_2: $100-a_{LL}$ $100-a_{HL}$

<u>A set of finite and pure equilibrium strategies</u> (n_{LL}, n_{LH}, n_{HL} odd, $n_{LL}\leq n_{LH}$, $n_{LL}\leq n_{HL}$)

H_1: *Demand and accept 99 before step n_{HL}, demand and accept a_{HL} in step n_{HL}, break off after step n_{HL}*

L_1: *Demand and accept 99 before step n_{LL}, demand and accept a_{LL} in step n_{LL}, demand and accept 99 between steps n_{LL} and n_{LH}, demand and accept a_{LH} in step n_{LH}, break off after step n_{LH}*

H_2: *Demand and accept 99 before step $n_{LH}+1$, demand and accept $100-a_{LH}$ in step $n_{LH}+1$, break off after $n_{LH}+1$*

L_2: *Demand and accept 99 before step $n_{LL}+1$, demand and accept $100-a_{LL}$ in step $n_{LL}+1$, demand and accept 99 between step $n_{LL}+1$ and $n_{HL}+1$, demand and accept $100-a_{HL}$ in step $n_{HL}+1$, break off after step $n_{HL}+1$*

This set of equilibrium strategies is constructed for n_{LL}, n_{LH}, and n_{HL} odd. This means that the first mover proposes the final agreement in all three cases. Analogously, like discussed in case 2, sets of equilibrium strategies can be constructed for all 8 possible combinations of n_{LL}, n_{LH}, and n_{HL} even and odd, respectively.

Furthermore, the strategies are constructed such that the two low type players agree before the high type and the low type on the equilibrium path. Obviously, a set of equilibrium strategies can also be constructed for the other cases.

For all admissible parameter constellations which satisfy the set of the necessary constraints this set of strategies is in equilibrium.

From inequalities (5.6) and (5.9) it follows that if $n_{LL} > n_{LH}$ and $n_{HL} > n_{HH}^*$ and n_{HH}^* even then

$$\delta_{HL} a_{HL} = \delta_{LL} a_{LL}. \tag{5.33}$$

If the agreement of L_1 and H_2 does not occur in n_{LH} (which is before n_{LL}), type L_1 knows that the opponent is weak. The strong type of player 1 knows that he plays a weak opponent in step n_{HH}^*, which is before n_{HL}. In order to exclude an imitation of the types of player 1 in the continuation of the game the two types of player 1 receive the same payoff.

From inequalities (5.7) and (5.10) it follows that if $n_{LL} > n_{HL}$ and $n_{LH} > n_{HH}^*$ and n_{HH}^* odd then

$$\delta_{LH}(100 - a_{LH}) = \delta_{LL}(100 - a_{LL}). \tag{5.34}$$

With analogous arguments (5.34) excludes an imitation of the two types of player 2 in case they can identify the weak type of player 1.

This leads to the expected payoffs:

$$P_H^1 = .5 \cdot (\delta_{HL} a_{HL} + \delta_{HH}^* \alpha) \tag{5.35}$$

$$P_H^2 = .5 \cdot (\delta_{LH}(100 - a_{LH}) + \delta_{HH}^* \alpha) \tag{5.36}$$

$$P_L^1 = .5 \cdot (\delta_{LH} a_{LH} + \delta_{LL} a_{LL}) \tag{5.37}$$

$$P_L^2 = .5 \cdot (\delta_{HL}(100 - a_{HL}) + \delta_{LL}(100 - a_{LL})) \tag{5.38}$$

<u>Case 6:</u> $p_{HH} = 1$, $p_{HL} = 1$, $p_{LH} = 1$, $p_{LL} = 1$

The set of the necessary constraints yields:

$$\delta_{HH}a_{HH} + \delta_{HL}a_{HL} \geq 2\alpha \tag{6.1}$$

$$\delta_{HH}(100-a_{HH}) + \delta_{LH}(100-a_{LH}) \geq 2\delta\alpha \tag{6.2}$$

$$\delta_{HH}a_{HH} + \delta_{HL}a_{HL} = \delta_{LH}a_{LH} + \delta_{LL}a_{LL} \tag{6.3}$$

$$\delta_{HH}(100-a_{HH}) + \delta_{LH}(100-a_{LH}) = \delta_{HL}(100-a_{HL}) + \delta_{LL}(100-a_{LL}) \tag{6.4}$$

$$\delta_{HH}a_{HH} \geq \delta_{HL}\delta\alpha, \text{ if } n_{HH} > n_{HL} \text{ and } n_{HL} \text{ even} \tag{6.5}$$

$$\delta_{HH}a_{HH} \geq \delta_{HL}\delta^2\alpha, \text{ if } n_{HH} > n_{HL} \text{ and } n_{HL} \text{ odd} \tag{6.6}$$

$$\delta_{HH}a_{HH} \geq \delta_{LH}a_{LH}, \text{ if } n_{HH} > n_{HL} \tag{6.7}$$

$$\delta_{HL}a_{HL} \geq \delta_{HH}\delta\alpha, \text{ if } n_{HH} < n_{HL} \text{ and } n_{HH} \text{ even} \tag{6.8}$$

$$\delta_{HL}a_{HL} \geq \delta_{HH}\delta^2\alpha, \text{ if } n_{HH} < n_{HL} \text{ and } n_{HH} \text{ odd} \tag{6.9}$$

$$\delta_{HL}a_{HL} \geq \delta_{LL}a_{LL}, \text{ if } n_{HH} < n_{HL} \tag{6.10}$$

$$\delta_{HH}(100-a_{HH}) \geq \delta_{LH}\delta\alpha, \text{ if } n_{HH} > n_{LH} \text{ and } n_{LH} \text{ odd} \tag{6.11}$$

$$\delta_{HH}(100-a_{HH}) \geq \delta_{LH}\delta^2\alpha, \text{ if } n_{HH} > n_{LH} \text{ and } n_{LH} \text{ even} \tag{6.12}$$

$$\delta_{HH}(100-a_{HH}) \geq \delta_{HL}(100-a_{HL}), \text{ if } n_{HH} > n_{LH} \tag{6.13}$$

$$\delta_{LH}(100-a_{LH}) \geq \delta_{HH}\delta\alpha, \text{ if } n_{HH} < n_{LH} \text{ and } n_{HH} \text{ odd} \tag{6.14}$$

$$\delta_{LH}(100-a_{LH}) \geq \delta_{HH}\delta^2\alpha, \text{ if } n_{HH} < n_{LH} \text{ and } n_{HH} \text{ even} \tag{6.15}$$

$$\delta_{LH}(100-a_{LH}) \geq \delta_{LL}(100-a_{LL}), \text{ if } n_{HH} < n_{LH} \tag{6.16}$$

$$\delta_{LH}a_{LH} \geq \delta_{HH}a_{HH}, \text{ if } n_{LL} < n_{LH} \tag{6.17}$$

$$\delta_{LL}a_{LL} \geq \delta_{HL}a_{HL}, \text{ if } n_{LL} > n_{LH} \tag{6.18}$$

$$\delta_{HL}(100-a_{HL}) \geq \delta_{HH}(100-a_{HH}), \text{ if } n_{LL} < n_{HL} \tag{6.19}$$

$$\delta_{LL}(100-a_{LL}) \geq \delta_{LH}(100-a_{LH}), \text{ if } n_{LL} > n_{HL} \tag{6.20}$$

$$a_{HL} \geq \alpha, \text{ if } n_{HL} \text{ odd} \tag{6.21}$$

$$a_{HL} \geq \delta\alpha, \text{ if } n_{HL} \text{ even} \tag{6.22}$$

$$100-a_{LH} \geq \alpha, \text{ if } n_{LH} \text{ even} \tag{6.23}$$

$$100-a_{LH} \geq \delta\alpha, \text{ if } n_{LH} \text{ odd} \tag{6.24}$$

$a_{HH} \geq \alpha$, if n_{HH} odd   (6.25)

$a_{HH} \geq \delta\alpha$, if n_{HH} even   (6.26)

$100-a_{HH} \geq \alpha$, if n_{HH} even   (6.27)

$100-a_{HH} \geq \delta\alpha$, if n_{HH} odd   (6.28)

$\delta_{LL}a_{LL} \geq 1$, if $n_{LL}=n_{LH}$ (6.29)

$\delta_{LH}a_{LH} \geq 1$, if $n_{LL}=n_{LH}$ (6.30)

$\delta_{LH}a_{LH} \geq \delta_{LL}$, if $n_{LL}<n_{LH}$ and n_{LL} even   (6.31)

$\delta_{LH}a_{LH} \geq \delta\delta_{LL}$, if $n_{LL}<n_{LH}$ and n_{LL} odd   (6.32)

$\delta_{LL}a_{LL} \geq \delta_{LH}$, if $n_{LL}>n_{LH}$ and n_{LH} even   (6.33)

$\delta_{LL}a_{LL} \geq \delta\delta_{LH}$, if $n_{LL}>n_{LH}$ and n_{LH} odd   (6.34)

$\delta_{LL}(100-a_{LL}) \geq 1$, if $n_{LL}=n_{HL}$ (6.35)

$\delta_{HL}(100-a_{HL}) \geq 1$, if $n_{LL}=n_{HL}$ (6.36)

$\delta_{HL}(100-a_{HL}) \geq \delta_{LL}$, if $n_{LL}<n_{HL}$ and n_{LL} odd   (6.37)

$\delta_{HL}(100-a_{HL}) \geq \delta\delta_{LL}$, if $n_{LL}<n_{HL}$ and n_{LL} even   (6.38)

$\delta_{LL}(100-a_{LL}) \geq \delta_{HL}$, if $n_{LL}>n_{HL}$ and n_{HL} odd   (6.39)

$\delta_{LL}(100-a_{LL}) \geq \delta\delta_{HL}$, if $n_{LL}>n_{HL}$ and n_{HL} even   (6.40)

From inequalities (6.25) and (6.26), as well as from inequalities (6.27) and (6.28) it follows

$$\alpha \leq \frac{100}{1+\delta}.$$ (6.41)

The following diagram shall illustrate an equilibrium path in a highly stylized way. Only the demands that yield the agreements are shown.

H_1: a_{HL} a_{HH}

L_1: a_{LL} a_{LH}

$\longrightarrow$

H_2: $100-a_{LH}$ $100-a_{HH}$

L_2: $100-a_{LL}$ $100-a_{HL}$

<u>A set of finite and pure equilibrium strategies</u> (n_{ij} odd for all i,j, $n_{LL} \leq n_{HL} \leq n_{LH} \leq n_{HH}$)

H_1:   *Demand and accept 99 before step n_{HL}, demand and accept a_{HL} in step n_{HL}, demand and accept 99 between steps n_{HL} and n_{HH}, demand and accept a_{HH} in step n_{HH}, break off after step n_{HH}*

L_1:   *Demand and accept 99 before step n_{LL}, demand and accept a_{LL} in step n_{LL}, demand and accept 99 between steps n_{LL} and n_{LH}, demand and accept a_{LH} in step n_{LH}, break off after step n_{LH}*

H_2:   *Demand and accept 99 before step $n_{LH}+1$, demand and accept $100-a_{LH}$ in step $n_{LH}+1$, demand and accept 99 between step $n_{LH}+1$ and $n_{HH}+1$, demand and accept $100 - a_{HH}$ in step $n_{HH}+1$, break off after step $n_{HH}+1$*

L_2:   *Demand and accept 99 before step $n_{LL}+1$, demand and accept $100-a_{LL}$ in step $n_{LL}+1$, demand and accept 99 between step $n_{LL}+1$ and $n_{HL}+1$, demand and accept $100-a_{HL}$ in step $n_{HL}+1$, break off after step $n_{HL}+1$*

This set of equilibrium strategies is constructed for n_{ij} odd, for all i,j. This means that the first mover proposes the final agreement in all cases. Analogously, like discussed in case 2, sets of equilibrium strategies can be constructed for all possible combinations of n_{LL}, n_{LH}, n_{HL}, and n_{HH} even and odd, respectively.

Furthermore, the equilibrium strategies can be constructed for all other relationships between the steps of agreement.

For all admissible parameter constellations which satisfy the set of the necessary constraints this set of strategies is in equilibrium.

From inequalities (6.7) and (6.17) it follows that if $n_{HH} > n_{HL}$ and $n_{LH} > n_{LL}$ then

$$\delta_{HH}a_{HH} = \delta_{LH}a_{LH}. \tag{6.42}$$

From inequalities (6.10) and (6.18) it follows that if $n_{HH} < n_{HL}$ and $n_{LH} < n_{LL}$ then

$$\delta_{HL}a_{HL} = \delta_{LL}a_{LL}. \tag{6.43}$$

From inequalities (6.13) and (6.19) it follows that if $n_{HH} > n_{LH}$ and $n_{HL} > n_{LL}$ then

$$\delta_{HH}(100-a_{HH}) = \delta_{HL}(100-a_{HL}). \tag{6.44}$$

From inequalities (6.16) and (6.20) it follows that if $n_{HH} < n_{LH}$ and $n_{HL} < n_{LL}$ then

$$\delta_{LH}(100-a_{LH}) = \delta_{LL}(100-a_{LL}). \tag{6.45}$$

The equalities of the payoffs exclude the incentive of an imitation of the other type after the player is able to distinguish between the two types of the opponent.

The expected payoffs in this case are:

$$P_H^1 = .5 \cdot (\delta_{HL}a_{HL} + \delta_{HH}a_{HH}) \tag{6.46}$$

$$P_H^2 = .5 \cdot (\delta_{LH}(100-a_{LH}) + \delta_{HH}(100-a_{HH})) \tag{6.47}$$

$$P_L^1 = .5 \cdot (\delta_{LH}a_{LH} + \delta_{LL}a_{LL}) \tag{6.48}$$

$$P_L^2 = .5 \cdot (\delta_{HL}(100-a_{HL}) + \delta_{LL}(100-a_{LL})) \tag{6.49}$$

We showed that for six (see table 3.2) of the 16 possible parameter constellations of agreement vectors p the set of the necessary constraints is not in contradiction. Furthermore, in each of these six cases a set of equilibrium strategies for each admissible parameter constellation satisfying the set of the necessary constraints could be constructed. This completes the proof of the theorem.   ◆

Remark:

Besides the case that the plays of all types end in conflict, two low types will always agree. On the other hand, two high types only agree if the plays of all types end in agreement. For the plays of a weak and a strong player all combinations of agreement probabilities are possible if the two low types agree and the two high types have a conflict.

3.2   THE CASE OF NO DISCOUNTING

As a special case we want to focus on the case of no discounting, this means $\delta=1$ and therefore $\delta_{ij}=\delta_{ij}^{*}=1$ for all $i,j \in \{H,L\}$. The assumption of no discounting simplifies the constraints and therefore allows us to gain a more lucid insight in the possible outcome configurations. Again, we shall look at the six different parameter constellations of the agreement probabilities belonging to Nash equilibria in pure and finite strategies. We shall present the necessary constraints and the expected payoffs in the simplified form for $\delta=1$. An outcome bimatrix will show the equilibrium outcomes. The outcome bimatrix contains four cells which contain the equilibrium outcomes for the four possible type combinations. Each cell of the outcome bimatrix contains a tuple, where the first component denotes the outcome of the first mover and the second component the outcome of the second mover.

<u>Case 1:</u> $p_{HH} = 0$, $p_{HL} = 0$, $p_{LH} = 0$, $p_{LL} = 0$

Second Mover

	H_2	L_2
H_1	α , α	α , 0
L_1	0 , α	0 , 0

First Mover

Since the expected payoffs do not depend on δ they do not change by choosing $\delta=1$.

$$P_H^1 = \alpha \tag{1.9}$$

$$P_H^2 = \alpha \tag{1.10}$$

$$P_L^1 = 0 \tag{1.11}$$

$$P_L^2 = 0 \tag{1.12}$$

<u>Case 2:</u> $p_{HH} = 0$, $p_{HL} = 0$, $p_{LH} = 0$, $p_{LL} = 1$

The constraints reduce to:

$$100 - \alpha \leq a_{LL} \leq \alpha \qquad (2.11)$$

which implies

$$\alpha \geq 50. \qquad (2.12)$$

Second Mover

		H_2	L_2
	H_1	α , α	α , 0
First Mover	L_1	0 , α	a_{LL} , $100 - a_{LL}$

The expected payoffs of the types emerge as

$$P_H^1 = \alpha \qquad (2.13)$$

$$P_H^2 = \alpha \qquad (2.14)$$

$$P_L^1 = .5a_{LL} \qquad (2.15)$$

$$P_L^2 = .5 \cdot (100 - a_{LL}) \qquad (2.16)$$

<u>Case 3:</u> $p_{HH} = 0$, $p_{HL} = 0$, $p_{LH} = 1$, $p_{LL} = 1$

The agreement outcomes have to fulfill:

$$a_{LL} = a_{LH} \qquad (3.18)$$

$$a_{LL} \leq \min\{\alpha, 100 - \alpha\}, \qquad (3.19)$$

which implies $a_{LL} \leq 50$.

Second Mover

	H_2	L_2
H_1	α , α	α , 0
L_1	a_{LL} , $100-a_{LL}$	a_{LL} , $100-a_{LL}$

First Mover

The expected payoffs are

$$P_H^1 = \alpha \tag{3.20}$$

$$P_H^2 = .5 \cdot (100 - a_{LL} + \alpha) \tag{3.21}$$

$$P_L^1 = a_{LL} \tag{3.22}$$

$$P_L^2 = .5 \cdot (100 - a_{LL}) \tag{3.23}$$

Case 4: $p_{HH} = 0$, $p_{HL} = 1$, $p_{LH} = 0$, $p_{LL} = 1$

The agreement outcomes have to satisfy:

$$a_{LL} = a_{HL} \tag{4.18}$$

$$a_{LL} \geq \max\{\alpha, 100-\alpha\}, \tag{4.19}$$

which implies $a_{LL} \geq 50$.

Second Mover

	H_2	L_2
H_1	α , α	a_{LL} , $100-a_{LL}$
L_1	0 , α	a_{LL} , $100-a_{LL}$

First Mover

The expected payoffs are

$$P_H^1 = .5 \cdot (a_{LL} + \alpha) \tag{4.20}$$

$$P_H^2 = \alpha \tag{4.21}$$

$$P_L^1 = .5 a_{LL} \tag{4.22}$$

$$P_L^2 = 100 - a_{LL} \tag{4.23}$$

<u>Case 5:</u> $p_{HH} = 0$, $p_{HL} = 1$, $p_{LH} = 1$, $p_{LL} = 1$

The set of restrictions reduces to:

$$a_{HL} \geq \alpha \tag{5.39}$$

$$a_{LH} \leq 100 - \alpha \tag{5.40}$$

$$a_{HL} \leq a_{LH} + a_{LL} \leq a_{HL} + \alpha \tag{5.41}$$

$$100 - a_{LH} \leq 100 - a_{HL} + 100 - a_{LL} \leq 100 - a_{LH} + \alpha \tag{5.42}$$

$$a_{LH} \leq \alpha, \text{ if } n_{HL} < n_{HH}^* \tag{5.43}$$

$$a_{HL} \geq a_{LL}, \text{ if } n_{HL} > n_{HH}^* \text{ and } n_{HH}^* \text{ even} \tag{5.44}$$

$$a_{LH} \leq a_{LL}, \text{ if } n_{LH} > n_{HH}^* \text{ and } n_{HH}^* \text{ odd} \tag{5.45}$$

$$100 - a_{HL} \leq \alpha, \text{ if } n_{LH} < n_{HH}^* \tag{5.46}$$

$$a_{LL} \geq a_{HL}, \text{ if } n_{LL} > n_{LH} \tag{5.47}$$

$$a_{LL} \leq a_{LH}, \text{ if } n_{LL} > n_{HL} \tag{5.48}$$

If $n_{LL} > n_{LH}$ and $n_{HL} > n_{HH}^*$ and n_{HH}^* even it follows from inequalities (5.44) and (5.47) that

$$a_{HL} = a_{LL}. \tag{5.49}$$

If $n_{LL} > n_{HL}$ and $n_{LH} > n_{HH}^*$ and n_{HH}^* odd it follows from inequalities (5.45) and (5.48) that

$$a_{LH} = a_{LL}. \tag{5.50}$$

Second Mover

	H_2	L_2
H_1	α , α	a_{HL} , $100-a_{HL}$
L_1	a_{LH} , $100-a_{LH}$	a_{LL} , $100-a_{LL}$

First Mover

This leads to the expected payoffs:

$$P_H^1 = .5 \cdot (a_{HL} + \alpha) \qquad (5.51)$$

$$P_H^2 = .5 \cdot (100 - a_{LH} + \alpha) \qquad (5.52)$$

$$P_L^1 = .5 \cdot (a_{LH} + a_{LL}) \qquad (5.53)$$

$$P_L^2 = .5 \cdot (100 - a_{HL} + 100 - a_{LL}) \qquad (5.54)$$

Case 6: $p_{HH} = 1$, $p_{HL} = 1$, $p_{LH} = 1$, $p_{LL} = 1$

In the case where all types agree, the set of the constraints reduces to:

$$a_{LL} = a_{LH} = a_{HL} = a_{HH} \qquad (6.50)$$

$$\alpha \le a_{LL} \le 100 - \alpha, \qquad (6.51)$$

which implies $\alpha \le 50$.

Second Mover

	H_2	L_2
H_1	a_{LL} , $100-a_{LL}$	a_{LL} , $100-a_{LL}$
L_1	a_{LL} , $100-a_{LL}$	a_{LL} , $100-a_{LL}$

First Mover

The expected payoffs in this case are:

$$P_H^1 = a_{LL} \tag{6.52}$$

$$P_H^2 = 100 - a_{LL} \tag{6.53}$$

$$P_L^1 = a_{LL} \tag{6.54}$$

$$P_L^2 = 100 - a_{LL} \tag{6.55}$$

In the case of the equilibria where an agreement is only reached in plays with two weak players ($p=(0,0,0,1)$) the inequality $\alpha \geq 50$ must hold, and in the case of the equilibria where the plays of all type combinations end with an agreement ($p=(1,1,1,1)$) the inequality $\alpha \leq 50$ must hold. For the other parameter constellations belonging to equilibrium points there are no restrictions on α which have to be satisfied for all parameter constellations.

3.3   THE EXPECTED EQUILIBRIUM PAYOFFS

Theorem 1 showed that a large number of equilibria exists in the two-person bargaining game with incomplete information. For six constellations of the agreement probabilities equilibria in pure and finite strategies could be constructed for each variation of the parameters in the admissible parameter range satisfying the set of the necessary constraints.

In a game playing experiment one cannot observe the underlying strategies of the subjects, but according to the payoffs which result one could examine whether it can be excluded that equilibrium strategies are played by the subjects. Therefore, we shall determine ranges in which the expected payoffs reached by equilibria in pure and finite strategies must be. These ranges are determined by the restrictions on the parameters and the constraints which could be deduced in each of the six cases of agreement probabilities in which equilibria in pure and finite strategies exist (see table 3.2). We shall distinguish $\delta = 1$ and $0 < \delta < 1$. In both cases we remain with the statement that a necessary condition for an expected equilibrium payoff is its location in the specified range. These ranges contain numbers which cannot be reached by a pure strategy equilibrium point of the two-person game with incomplete information, but no expected equilibrium payoff is located outside a range.

Notation:

Let P_L^1 be the expected payoff of a first mover low type in a pure strategy equilibrium point for a two-person bargaining game with incomplete information. Let P_L^2 be the expected payoff of a second mover low type in a pure strategy equilibrium point for a two-person bargaining game with incomplete information. Similarly, let P_H^1 be the expected payoff of a first mover high type in a pure strategy equilibrium point for a two-person bargaining game with incomplete information and let P_H^2 be the expected payoff of a second mover high type in a pure strategy equilibrium point for a two-person bargaining game with incomplete information.

Theorem 2:

Let $\delta=1$. Then the following is true for the expected payoffs in a pure strategy equilibrium point for a two-person bargaining game with incomplete information.

$$
P_L^i \in \begin{cases} [0,100-\alpha], & \text{for } 1\leq\alpha\leq 33\frac{2}{3} \\[2ex] [0,49.5+\frac{\alpha}{2}], & \text{for } 33\frac{2}{3}<\alpha\leq 50 \quad i=1,2 \\[2ex] [0,99.5-\frac{\alpha}{2}], & \text{for } 50<\alpha<100 \end{cases}
$$

$$
P_H^i \in \begin{cases} [\alpha,100-\alpha], & \text{for } 1\leq\alpha\leq 33\frac{2}{3} \\[2ex] [\alpha,49.5+\frac{\alpha}{2}], & \text{for } 33\frac{2}{3}<\alpha<100 \end{cases} \quad i=1,2
$$

Proof:

In six different cases of agreement probabilities it was possible to construct equilibria in pure and finite strategies for the two-person bargaining game with incomplete information (see Section 3.1 and especially table 3.2). For each of these six cases we shall determine the expected equilibrium payoffs for the four types of players (P_H^1, P_H^2, P_L^1, and P_L^2). The necessary conditions together with admissible parameter ranges specify a range for these expected equilibrium payoffs. The minimal lower bound and the maximal upper bound over the six cases are the lower and the upper bound of the range of the expected equilibrium payoffs. We shall see that the ranges of the first and the second mover of the low type coincide and that the ranges of the first and the second mover of the high type coincide.

The proof distinguishes among the four types of players and $\alpha\leq 50$ and $\alpha>50$. Remember, that case 6 requires $\alpha\leq 50$ and that case 2 requires $\alpha>50$.

Let us first consider the low types. Let $\alpha\leq 50$.

Case 1: $P_L^1 = 0$ $P_L^2 = 0$

Case 3: $1 \leq P_L^1 = a_{LL} \leq \alpha$ $\frac{1}{2}(100-\alpha) \leq P_L^2 = \frac{1}{2}(100-a_{LL}) \leq 49.5$

Case 4: $\frac{1}{2}(100-\alpha) \leq P_L^1 = \frac{1}{2}a_{LL} \leq 49.5$ $1 \leq P_L^2 = 100-a_{LL} \leq \alpha$

Case 5:

$\frac{1}{2}\alpha \leq P_L^1 = \frac{1}{2}(a_{LH}+a_{LL}) \leq \frac{1}{2}(a_{HL}+\alpha) \leq 49.5+\frac{1}{2}\alpha$ according to inequality (5.41)

on the other hand: $P_L^1 = \frac{1}{2}(a_{LH}+a_{LL}) \leq \frac{1}{2}(100-\alpha+99) = 99.5-\frac{1}{2}\alpha$

$\Rightarrow P_L^1 \leq \min\{49.5+\frac{1}{2}\alpha, 99.5-\frac{1}{2}\alpha\}$

$\frac{1}{2}\alpha \leq P_L^2 = (100-a_{HL}+100-a_{LL}) \leq \frac{1}{2}(100-a_{LH}+\alpha) \leq 49.5+\frac{1}{2}\alpha$ according to (5.42)

on the other hand: $P_L^2 = (100-a_{HL}+100-a_{LL}) \leq \frac{1}{2}(100-\alpha+99) = 99.5-\frac{1}{2}\alpha$

$\Rightarrow P_L^2 \leq \min\{49.5+\frac{1}{2}\alpha, 99.5-\frac{1}{2}\alpha\}$

Case 6: $\alpha \leq P_L^1 = a_{LL} \leq 100-\alpha$ $\alpha \leq P_L^2 = 100-a_{LL} \leq 100-\alpha$

Let $\alpha > 50$.

Case 1: $P_L^1 = 0$ $P_L^2 = 0$

Case 2: $50-\frac{1}{2}\alpha \leq P_L^1 = \frac{1}{2}a_{LL} \leq \frac{1}{2}\alpha$ $50-\frac{1}{2}\alpha \leq P_L^2 = \frac{1}{2}(100-a_{LL}) \leq \frac{1}{2}\alpha$

Case 3: $1 \leq P_L^1 = a_{LL} \leq 100-\alpha$ $\frac{1}{2}\alpha \leq P_L^2 = \frac{1}{2}(100-a_{LL}) \leq 49.5$

Case 4: $\frac{1}{2}\alpha \leq P_L^1 = \frac{1}{2}a_{LL} \leq 49.5$ $1 \leq P_L^2 = 100-a_{LL} \leq 100-\alpha$

Case 5:

$\frac{1}{2}\alpha \leq P_L^1 = \frac{1}{2}(a_{LH}+a_{LL}) \leq \frac{1}{2}(a_{HL}+\alpha) \leq 49.5+\frac{1}{2}\alpha$ according to inequality (5.41)

on the other hand: $P_L^1 = \frac{1}{2}(a_{LH}+a_{LL}) \leq \frac{1}{2}(100-\alpha+99) = 99.5-\frac{1}{2}\alpha$

$\Rightarrow P_L^1 \leq \min\{49.5+\frac{1}{2}\alpha, 99.5-\frac{1}{2}\alpha\}$

$\frac{1}{2}\alpha \leq P_L^2 = (100-a_{HL}+100-a_{LL}) \leq \frac{1}{2}(100-a_{LH}+\alpha) \leq 49.5+\frac{1}{2}\alpha$ according to (5.42)

on the other hand: $P_L^2 = (100-a_{HL}+100-a_{LL}) \leq \frac{1}{2}(100-\alpha+99) = 99.5-\frac{1}{2}\alpha$

$\Rightarrow P_L^2 \leq \min\{49.5+\frac{1}{2}\alpha, 99.5-\frac{1}{2}\alpha\}$

Now consider the two high type players. Let $\alpha \leq 50$.

Case 1: $P_H^1 = \alpha$ $P_H^2 = \alpha$

Case 3: $P_H^1 = \alpha$ $50 \leq P_H^2 = \frac{1}{2}(100-a_{LL}+\alpha) \leq 49.5+\frac{1}{2}\alpha$

Case 4: $50 \leq P_H^1 = \frac{1}{2}(a_{LL}+\alpha) \leq 49.5+\frac{1}{2}\alpha$ $P_H^2 = \alpha$

Case 5: $\alpha \leq P_H^1 = \frac{1}{2}(a_{HL}+\alpha) \leq 49.5+\frac{1}{2}\alpha$

$\alpha \leq P_H^2 = \frac{1}{2}(100-a_{LH}+\alpha) \leq 49.5+\frac{1}{2}\alpha$

Case 6: $\alpha \leq P_H^1 = a_{LL} \leq 100 - \alpha$ $\alpha \leq P_H^2 = 100 - a_{LL} \leq 100 - \alpha$

Let $\alpha > 50$.

Case 1: $P_H^1 = \alpha$ $P_H^2 = \alpha$

Case 2: $P_H^1 = \alpha$ $P_H^2 = \alpha$

Case 3: $P_H^1 = \alpha$ $\alpha \leq P_H^2 = \frac{1}{2}(100 - a_{LL} + \alpha) \leq 49.5 + \frac{1}{2}\alpha$

Case 4: $\alpha \leq P_H^1 = \frac{1}{2}(a_{LL} + \alpha) \leq 49.5 + \frac{1}{2}\alpha$ $P_H^2 = \alpha$

Case 5: $\alpha \leq P_H^1 = \frac{1}{2}(a_{HL} + \alpha) \leq 49.5 + \frac{1}{2}\alpha$

$$\alpha \leq P_H^2 = \frac{1}{2}(100 - a_{LH} + \alpha) \leq 49.5 + \frac{1}{2}\alpha$$

For the determination of the maximal upper bound of the low type's expected equilibrium payoff we have to distinguish between $1 \leq \alpha \leq 33\frac{2}{3}$, $33\frac{2}{3} < \alpha \leq 50$, and $50 < \alpha < 100$. For the maximal upper bound of the high type's expected equilibrium payoff we have to distinguish between $1 \leq \alpha \leq 33\frac{2}{3}$, and $33\frac{2}{3} < \alpha < 100$. In the first case the upper bound $100 - \alpha$ is greater than the upper bound $49.5 + \frac{1}{2}\alpha$ and in the latter case the converse is true. Notice, that the value 49.5 occurs as $\frac{1}{2}99$. The value 99 is the upper bound of an outcome. This leads to the ranges as formulated in the theorem.   ◆

Remarks:

(I)   Theorem 2 showed that the first mover and the second mover of a type (H or L) are not distinguished concerning the range of the expected equilibrium payoffs. Since $\delta = 1$, the mover sequence does not affect the expected payoffs.

(II)   Theorem 2 showed that every expected payoff of a pure strategy equilibrium point of the two-person bargaining game with incomplete information is located in the above specified ranges. This induces the question whether each point of the above specified ranges is an expected payoff of a pure strategy equilibrium point of the two-person bargaining game with incomplete information. Except for case 5 this question can be answered easily. Consider an arbitrary case, but not case 5. The necessary constraints on the agreement outcomes do not depend on the steps of agreement n_{ij}. Since the expected equilibrium payoffs are either of the form a_{ij} or $\frac{1}{2}a_{ij}$, all integer numbers or halves of integer numbers in the specified range are exactly the expected equilibrium payoffs. In case 5 this is more complicated. Here we have restrictions which specify relationships between the agreement outcomes contingent on the step of agreement. This means that dependent on the step of agreement we might not be able to reach exactly each integer or halve of an integer as an expected equilibrium payoff. Due to this problem we shall refrain from exactly specifying the set of the expected equilibrium payoffs.

For the case of discounting $(0 < \delta < 1)$ the same considerations as in theorem 2 can be made to deduce the bounds for the expected equilibrium payoffs of the four types of players.

Theorem 3:

Let $0 < \delta < 1$. Then the following is true for the expected payoffs in a pure strategy equilibrium point for a two-person bargaining game with incomplete information.

$$P_L^1 \in \begin{cases} [0, 99.5 - \frac{\alpha}{2}], & \text{for } 1 \leq \alpha \leq \frac{100}{1+\delta} \\ [0, \min\{49.5 + \frac{\alpha}{2}, 99.5 - \frac{\alpha}{2}\}], & \text{for } \frac{100}{1+\delta} < \alpha < 100 \end{cases}$$

$$P_L^2 \in \begin{cases} [0, 99.5 - \frac{\alpha}{2}], & \text{for } 1 \leq \alpha \leq \frac{100}{1+\delta} \\ [0, \min\{49.5 + \frac{\alpha}{2}, 99.5 - \frac{\alpha}{2}\}], & \text{for } \frac{100}{1+\delta} < \alpha < 100 \end{cases}$$

$$P_H^1 \in \begin{cases} [\alpha, \max\{99.5 - \frac{\alpha}{2}, 49.5 + \frac{\alpha}{2}\}], & \text{for } 1 \leq \alpha \leq \frac{100}{1+\delta} \\ [\alpha, 49.5 + \frac{\alpha}{2}], & \text{for } \frac{100}{1+\delta} < \alpha < 100 \end{cases}$$

$$P_H^2 \in \begin{cases} [\delta\alpha, \max\{99.5 - \frac{\alpha}{2}, 49.5 + \frac{\alpha}{2}\}], & \text{for } 1 \leq \alpha \leq \frac{100}{1+\delta} \\ [\delta\alpha, 49.5 + \frac{\alpha}{2}], & \text{for } \frac{100}{1+\delta} < \alpha < 100 \end{cases}$$

<u>Proof:</u>

The ranges of the expected equilibrium payoffs are determined by the necessary constraints and the admissible parameter ranges deduced in Section 3.1. These constraints allow to deduce lower and upper bounds for the expected equilibrium payoffs. We shall refrain from explicitly listing the calculations to deduce the bounds. These calculations are very similar to those of theorem 2. The resulting bounds will be summarized in tables 3.3 and 3.4. The tables distinguish the six different cases of agreement probabilities for which equilibria in pure and finite strategies exist (see table 3.2). Remember that case 6 requires $\alpha \leq 100/(1+\delta)$.

<u>Table 3.3:</u> Lower and upper bounds of the expected equilibrium payoffs of the low type players for $0 < \delta < 1$

Case	Bounds for P_L^1		Bounds for P_L^2	
	lower	upper	lower	upper
1	0	0	0	0
2	$\frac{1}{2}$	$\frac{1}{2}(2-\delta)\alpha$	$\frac{1}{2}$	$\frac{1}{2}\delta\alpha$
3	0	α	$\frac{1}{2}\delta\alpha$	49.5
4	$\frac{1}{2}(2-\delta)\alpha$	49.5	0	$\delta\alpha$
5	$\frac{1}{2}\alpha$	$\min\{49.5+\frac{1}{2}\alpha, 99.5-\frac{1}{2}\alpha\}$	$\frac{1}{2}(2\delta-1)\alpha$	$\min\{49.5+\frac{1}{2}\alpha, 99.5-\frac{1}{2}\alpha\}$
6	α	$99.5-\frac{1}{2}\alpha$	$\delta\alpha$	$99.5-\frac{1}{2}\alpha$

<u>Table 3.4:</u> Lower and upper bounds of the expected equilibrium payoffs of the high type players for $0 < \delta < 1$

Case	Bounds for P_H^1		Bounds for P_H^2	
	lower	upper	lower	upper
1	α	α	α	α
2	α	α	$\frac{1}{2}(1+\delta)\alpha$	$\frac{1}{2}(1+\delta)\alpha$
3	α	α	$\frac{1}{2}(1+\delta)\alpha$	$49.5+\frac{1}{2}\alpha$
4	α	$49.5+\frac{1}{2}\delta\alpha$	$\delta\alpha$	$\delta\alpha$
5	α	$49.5+\frac{1}{2}\alpha$	$\delta\alpha$	$49.5+\frac{1}{2}\alpha$
6	α	$99.5-\frac{1}{2}\alpha$	$\delta\alpha$	$99.5-\frac{1}{2}\alpha$

For each type the minimal lower bound and the maximal upper bound determine the bounds of the range of the expected equilibrium payoffs. Case 6 makes it necessary to distinguish $\alpha \leq 100/(1+\delta)$ and $\alpha > 100/(1+\delta)$. This leads to the ranges stated in the theorem.   ◆

Remark:

Like in the case of $\delta = 1$ the location of an expected payoff in the specified ranges is only a necessary condition. Due to the set of the necessary conditions and the admissible parameter ranges there are elements of the specified ranges which are no expected equilibrium payoffs. But, we will find no expected equilibrium payoff outside these ranges.

In Section 6.7 we shall compare the average payoffs of the game playing experiment with the ranges of the expected equilibrium payoffs, deduced in this section. Section 12.3 will study the expected payoffs in the strategy experiment. For these investigations the specification of the ranges as the necessary condition for an expected equilibrium payoff suffices.

CHAPTER 4. RELATED BARGAINING EXPERIMENTS

4.1   TWO-PERSON BARGAINING EXPERIMENTS WITH INCOMPLETE INFORMATION

In the experimental literature there is only a small number of investigations on game situations which involve incomplete information. Although they are all different from our experiment we shall very briefly report their findings in order to analyze similarities to ours.

A bargaining experiment involving incomplete information, which is very close to ours, was performed by Hoggatt, Selten, Crockett, Gill and Moore (1978). This experiment is based on the theory of bargaining under incomplete information by Harsanyi and Selten (1972) and examines the example analyzed in detail in Selten (1975).

The underlying bargaining problem of the experiment by Hoggatt et al. (1978) is the following. Two players bargain over the division of 20 money units. In case of conflict both players receive nothing. In case of agreement they receive their agreement payoff minus a private cost. The cost can be either high (H) or low (L). A high cost is 9, while a low cost is 0. The private cost of each player is chosen randomly before the experiment, with high and low cost having equal probability. The random draws for both players are independent. Each player is aware of his own cost, but is ignorant about the other player's cost. The probability distribution and the fact that the costs of both players are drawn from this distribution are common knowledge. Time is divided into stages. At each stage the two players simultaneously demand a share of 20, which has to be not smaller than the private cost and smaller than 20. Furthermore, the demand of a player is not allowed to exceed the demand of the previous stage. If both players repeat their demand of the previous stage, the bargaining ends in conflict. If they demand values which sum up to at most 20, each player receives his demand plus half of the difference between the sum of the demands and 20, minus his private cost. In this case the game ends in agreement. In the case that at least one of both players does not repeat his demand and the sum of the demands exceeds 20, the bargaining proceeds to the next stage.

The game theoretic solution of this game (*generalized Nash solution*) is developed by Harsanyi and Selten (1972) and calculated by Selten (1975) for this particular example. Selten calculates an approximate solution which can be represented by a probability mixture of at most two strict equilibrium points in pure strategies, called the *main representation*. The main representation prescribes that a type H player should always demand 14. A player of type L should demand 14 in step 1, 10 in step 2, and 6 in step 3. This means that two players of type H always fail to agree, a player of type H agrees with a player of type L on (14,6) and two type L players agree on (10,10). Remember, that a type H player has to pay a cost of 9 in case of agreement.

The data of the experiment show the following conflict frequencies: 0.729 for (H,H) plays, 0.471 for (H,L) plays, and 0.097 for (L,L) plays. Almost all agreements between two high type players were equal split agreements. The agreement payoffs show a similarity in the distribution for (L,H) plays and (L,L) plays (the Kolmogoroff-Smirnov-test cannot significantly distinguish them). The theory predicts that two L type players should divide the 20 equally. Actually, only about one quarter of all agreements are equal splits. Hoggatt et al. investigate whether this asymmetry is due to a "bluffing" of one type L player. But, they found that in about 60% the L type player with the higher final payoff did not repeat his demand even once. They state that unequal payoffs mostly do not occur because one player is a "hard" bargainer, but because one player is a "soft" bargainer (which means that he had a lower initial demand or he lowered his demands more quickly). An operational definition of playing *weak* or *soft* is given over the initial demands and the concession rate. A player is seen to be weak if his initial demand is lower than 15 or if his demand change from step to step is greater than 1. Recall, that a demand change of 1 would be sufficient in order to avoid a conflict. Since each player played 5 games in succession, Hoggatt et al. were able to observe the behavior of playing weak over time. They found that with greater experience the players learn to avoid weak moves. A consequence of this learning behavior would be that the number of rounds played increases (due to tougher bargaining). Indeed, Hoggatt et al. found that the average bargaining length increased strictly monotonically from 4.974 in the first game to 8.812 in the fifth game.

In a series of three experiments Roth et al. study the influences of information on binary lottery games (Roth and Malouf, 1979, Roth, Malouf and Murnighan, 1981, and Roth and Murnighan, 1982). A *binary lottery game* is a game in which each player has two possible payoffs which occur with complementary probabilities. The higher payoff corresponds to winning the lottery, the other one to loosing it. It is possible that the monetary payoffs of the lotteries differ for the two players. The two players bargain over the probability of winning the (personal) lottery. Specifically, they bargain over the allocation of 100 lottery tickets. Receiving x lottery tickets corresponds to a chance of x% of winning the lottery.

The three above mentioned experiments differ in the provision of the information on the opponent's lottery and whether the information structure of the game is common knowledge. Furthermore, a currency "chips" is invented and the design varies in the information on the exchange rate of this currency. Very briefly, and for our purpose most interestingly, the results of the three experimental investigations can be summarized as follows: "the outcomes in the partial information condition (each player knows only his own lottery payoffs) tended to be extremely close to an equal split division of the lottery tickets, while outcomes in the full information condition (each player knows both lottery payoffs) shifted towards equal expected monetary payoffs". Roth and Murnighan (1982) give the following explanation. "In conflicts involving a wide range of potential agreements, social conventions may serve to make some agreements and demands more credible than other ones. Thus this hypothesis views the low variance observed in the partial (low) information condition as evidence that the agreement giving players an equal chance of winning the prizes is supported by a social norm that inclines both players to believe that their opponent may not accept less. The shift towards equal expected monetary payoffs observed in the full information condition is viewed as evidence that when information about the monetary value of the prize is available, the agreement giving the players equal expected payoffs is also supported by such a convention, so the bargaining focuses on resolving the difference between two credible positions." By *social convention* Roth et al. mean "customs or beliefs which are commonly shared by the members of a particular society".

Mitzkewitz and Nagel (1993) examined an ultimatum bargaining game under incomplete information. There, only the allocator is aware of the size of the cake to distribute. The receiver only knows the probability distribution over the cake sizes. In one version of the game the allocator *demands* a share for himself, in the other version the allocator *offers* a share to the receiver. In both versions the receiver can reject (and both players receive nothing) or accept. In the *demand game* the receiver does not know his payoff in case of acceptance (he receives the residual from the cake size and the allocator's demand), whereas in the *offer game* he knows his own payoff (the offer), but not the payoff of the allocator (which is the residual from the cake size and the offer).

Mitzkewitz and Nagel compared the experimental results with the 50-50-split and the sequential equilibrium prediction. They found that none of these concepts alone has predictive power but constructed a descriptive theory, the *anticipation strategy*, mainly based on these concepts. The idea behind the anticipation strategy is that the allocator forms expectations about the receiver's acceptance level, which are based on fairness considerations. It is assumed that the receiver is willing to punish visible unfairness. The allocator then chooses a best reply against the expected behavior of the receiver. The predictive power of the anticipation strategy is found to be much greater than that of the 50-50-strategy or the equilibrium strategy. However, Mitzkewitz and Nagel found that the 50-50-rule has relevance for inexperienced subjects, but it diminishes over time.

Croson (1992) examines four different designs of ultimatum bargaining games, where the offer structure and the information structure were varied. The allocator offers either a fixed dollar amount or a percentage of the cake. Either only the allocator or both players have knowledge of the cake size. She conducted experiments under all four different conditions and concluded, that "proposers offer less and responders claim they will accept less in ultimatum games when the size of the pie is unknown to the responders and the offers are made in dollar amounts. Additionally, responders claim they demand much more when the offer is in percentages and they are informed about the size of the pie than in any other treatment".

A recent paper by Kennan and Wilson (1993) gives a broad overview over bargaining problems with incomplete information, in the theoretical as well as in the experimental literature.

4.2   FAIRNESS IN BARGAINING

There is a variety in recent experimental research that addresses the topic of subject's conception of fairness. Hoffman and Spitzer (1982 and 1985) distinguish three different types of concepts: the *utilitarian (economic man)* who makes only selfregarding decisions; the *egalitarian* who strives for equal shares, and the behavior according to the *natural law/desert* theory, which asserts that, as a matter of natural law, someone deserves resources. In the earlier experiment they studied a two-person bargaining game where one player (the *controller*) could either take $12 or enter a bargaining with the other player about $14. All controllers decided to enter the bargaining game which ended with the equal split allocation. The authors summarize the motivation of the players as: "because the players were morally equal, an equal split seemed to be the only fair allocation". In the later experiment they compared the effects of assigning the role of the controller randomly or earning the role from winning a preliminary game (hash game). Moreover, they investigated the effects of the difference in the presentation of the role assignment. In the *moral authority* setup subjects were told that they "earned the right of being the controller" while in the *no moral authority* setup they were told that they are "designated" to be the controller. The findings of the experiment are that subjects behave according to the *natural law/desert* instead of the *egalitarism* or *utilitarism*. This means winning the preplay is seen as a justified right to receive a larger payoff than the other player. In the case where the position of the controller is assigned by a flip of the coin, the payoffs are more egalitarian. The *moral authority* setup strengthens the results.

The influences of moral authority on fairness are studied in Hoffman, McCabe, Shachat and Smith (1992). They designed an experimental setup for a dictator game that ensured anonymity with respect to the experimenter besides the between-subjects anonymity in order to minimize the influence of the subjects' beliefs about the experimenters objectives (*double blind*). In this experiment the most "greedy" behavior of the controller is observed in comparison to previous experiments on dictator and ultimatum bargaining games. These experiments suggest that subjects behave more egalitarian if they did not "earn" their position in the game and are non-anonymous with respect to the experimenter. As soon as entitlement or double blindness is introduced the subjects tend to act more greedily.

Kahneman et al. (1986b) investigate the problem of fairness in market price exchanges. They propose the *principle of dual entitlements*, which states that a price increase is judged as fair by the consumers if they belief that it is due to an increase in the producers costs. A laboratory experiment conducted by Kachelmaier et al. (1991) supports the principle of dual entitlements by finding that higher market prices emerge in markets where the consumers are informed about the increase in costs than in control markets with ignorant consumers.

Kahneman et al. (1986b) state that "the cardinal rule of fair behavior is surely that one person should not achieve a gain by simply imposing an equivalent loss on another". A two-stage bargaining game conducted by Kahneman et al. (1986a) provides evidence that the bargaining behavior is inconsistent with the assumption that players are motivated solely by monetary rewards. The authors found a "willingness to pay for justice" by "punishing an unfair allocator". Players who did not divide the prize equally in the first stage of the game were "punished" by lower payoffs (than egalitarians) in the second stage of the game, even though the allocator had a slightly lower payoff caused by the punishment.

There is a large number of investigations in fairness in ultimatum bargaining games (which will only be discussed very briefly). The controversy whether fairness considerations play a role started with the different findings by Güth, Schmittberger and Schwartz (1982) and Binmore, Shaked and Sutton (1985). Güth et al. found that in a (one-stage) ultimatum game the average proposal of the allocator was 65% for inexperienced and 69% for experienced players, far away, however, from the subgame perfect equilibrium prediction of 100% or 100% minus the smallest money unit. Nevertheless, Binmore et al. found in a two-stage ultimatum game, where the pie was discounted by 75% in the second round, that 37% of the experienced players propose the equilibrium demand. Güth and Tietz (1988) suggested that the division according to the subgame perfect equilibrium prediction in the experiments of Binmore et al. (1985) is not so extreme, such that fairness consideration will not displace strategic considerations. They replicated the experiment by Binmore et al. (1985) with a high discount factor for the pie, leading to an extreme division according to the equilibrium, and found average first demands of 70% for inexperienced players and 67% for experienced players. Prasnikar and Roth (1992) added an experiment on a market game with a very extreme distribution prediction (by the subgame perfect equilibrium) to this discussion. They experimented a market with nine sellers and one buyer. Every buyer has a redemption value of $10 and bids for buying the good from the seller. The (extreme) subgame perfect equilibrium division is that the buyers are willing to pay $10 or $10 minus the smallest money unit. This solution emerges in the data of experienced players. It is the author's opinion that the emergence of the equilibrium allocation is due to the competitive situation among the buyers, which cannot directly be compared to the allocation of an amount of money between two players.

Ochs and Roth (1989) drew, from the study of previous ultimatum experiments and the observation of additional ones, the conclusion that "bargaining is a complex social phenomenon, which gives bargainers systematic motivations distinct from simple maximization". They propose to incorporate distributional considerations into the bargainer's utility function.

And, Thaler (1988) states, that "most people prefer more money to less, like to be treated fairly, and like to treat others fairly".

The idea by Ochs and Roth (1989) was taken up by Bolton (1991). In his *comparative model of bargaining* Bolton incorporates fairness into the utility function of the bargainers. In a two stage ultimatum bargaining experiment he observed that bargainers are concerned about the absolute as well as the relative payoff. In the first stage of the considered game player α offers an allocation of the coalition value. The second player β can either accept (and the game ends with the allocation proposed by α) or enter a second round, where β is the proposer and the coalition value is discounted (with different discount rates for both players). In the second round α can either accept or the game ends with both players receiving zero payoffs. A large number of disadvantageous counteroffers by β can be observed (like in the study of Ochs and Roth, 1989). Bolton explains these occurrences by stating that besides the money payoff the players are interested in the relation to the other player's payoff: "bargainers are trading away absolute money in order to gain relative money". In consequence he proposes a model in which fairness (relative money) is a component of the utility function. The subgame perfect equilibria of this comparative model of bargaining perform well in explaining the observed behavior in the experiment. Furthermore, Bolton shows that in a *tournament setup* where the payoff of an α player is determined by his payoff ranking under all α players in the experiment, the number of disadvantageous offers diminishes.

Yaari and Bar-Hillel (1984) discuss the notion of justice from an ethical viewpoint.

CHAPTER 5. THE EXPERIMENTAL DESIGN

5.1   TECHNICAL CONDITIONS

The experiment was conducted at the Bonn Laboratory of Experimental Economics. The subjects were mostly students of economics and law who never participated in a two-person bargaining game before. They were informed about the bargaining rules in a 20 minutes introductory session (for details, see Appendix A). This introduction also provided the information about the point to cash rate and the subjects were told that their objective should be the maximization of their payoffs. Afterwards they were seated in separate cubicles in the laboratory. Each cubicle was equipped with a computer terminal which was connected via a network to the other terminals. The interaction of the subjects was controlled by the terminal program. The bargaining was anonymous, which means that a subject neither knew the name of the opponent nor the cubicle he was seated in. The communication between the subjects was restricted to the formal interactions of proposing, accepting, and breaking off. No verbal communication was permitted. The information provided on the computer screen consisted of the alternative of the player, the coalition value, and the complete history of the bargaining process. The subjects had no access to information about games in which they were not participants.

Immediately after the session the subjects were paid individually and privately according to their success. Each bargaining point was exchanged by the point to cash rate into money. Cash was the only incentive offered to the subjects.

A session lasted about three to four hours and the point to cash rate was calculated such that a player gained on average 10 to 12 Deutsch Marks per hour, which is equivalent to an hour's work in a student's next-best alternative employment.

The value of the high alternative α was fixed in each session, and for each player an initial random draw decided whether he had the high alternative value α or the low alternative value 0. Moreover, the first decider was determined randomly with equal probability for both players.

5.2   THE EXPERIMENTAL DESIGN

The game theoretic analysis of the two-person bargaining game with incomplete information, introduced in Section 2.1, is very difficult, and the large variety of Nash equilibria of this game makes it difficult to gain a clear picture of the outcomes. An experiment based on this game should show how subjects behave in the bargaining situation.

The two-person bargaining game is experimentally tested for five different parameter values of the high alternative value α: 30, 45, 55, 60, and 70. For $\alpha=30$ and $\alpha=45$ the game is superadditive for all four type combinations. For the values of $\alpha>50$ two high type players cannot reach an agreement yielding individually rational outcomes for both players. The parameters for α were chosen to cover a large variety of the strategic aspects of the game. The values 45 and 55 are closely below and above half of the coalition value, and the value 30 shall represent a low value of the high alternative. A high alternative value of 70 puts the high type player in a very strong position by enabling him to receive 70% of the coalition value by the unilateral decision of a break off. The value of $\alpha=60$ is chosen as an intermediate high alternative value exceeding 50. It is not seen as meaningful to exceed 70 as the high alternative value, since this may terminate the bargaining immediately by break offs by the strong player.

We chose a discount factor of $\delta=.99$. Such a low discounting is viewed as most realistic.

All sessions involved a smallest money unit of $\mu=1$. The players were allowed to demand all integer values in the range $[0,100]$. This deviates from the game considered in the game theoretic analysis. There, we considered a game where the players were allowed to demand all integer values in the range $[1,99]$. In the game theoretic analysis this range was chosen in order to simplify the analysis, since in this case one has not to be concerned about a low type's indifference between a conflict outcome of 0 and a proposal of 0. However, it is my opinion that this difference in the demand ranges is, from a behavioral point of view, of minor importance. A demand of 100 was, in the same way as 99, sometimes chosen as a "threat". But, there was no subject which demanded 99 or 100 for a longer time period and no agreement occurred at 99 or 100.

For each parameter value of α (treatment) six sessions were conducted, such that we gained data from 30 sessions of the two-person bargaining game with incomplete information. Each session is an independent subject group and therefore a statistically independent observation. In total we observed 1440 plays of the two-person bargaining game with incomplete information.

The subject group of each session contained six subjects and each subject played 16 games with anonymous opponents of the group, changing from play to play. These 16 games consisted of four successions of a random order of the four game situations a player can be in. These four *game situations* are: (L,L), (L,H), (H,L), and (H,H), where the first entry of the pair denotes the type of the player and the second entry the type of the opponent. Clearly, from the players point of view situations 1 and 2, and situations 3 and 4 are not distinguishable. Hence, in this setup the types of the players were not drawn randomly but predetermined by the experimenter in a way that did not contradict the information on the probability distribution, given to the subjects. This predetermination was chosen in order to have an equal number of observations for each game situation and to have four levels of experience. In the first level, the players experience each situation for the first time, in the second level they experience each situation for the second time, and so on. The subjects were ignorant about this special design of the setup; they were told that the alternative values were drawn randomly with equal and independent probabilities. Table 5.1 shows the succession of the games in each session.

From game to game the opponent of a player changed. Since there were only five possible opponents, a player met each opponent at least three times and one opponent four times. But, a player met an opponent for the second time only if all other possible pairs of players have met for the first time, and so on.

After each play a subject was explicitly informed about the payoff (discounted outcome) of this play on the computer screen (see Appendix A). An information about the opponent's type was not provided.

The point to cash rate was 1 to .06 Deutsch Marks in each session of the experiment.

Table 5.1: Succession of games in each session

Game Nº	Alternative value player	Alternative value opponent
1	0	0
2	0	α
3	α	0
4	α	α
5	α	0
6	α	α
7	0	α
8	0	0
9	α	α
10	α	0
11	0	0
12	0	α
13	α	0
14	0	0
15	α	α
16	0	α

5.3   Notes on the Evaluation of the Experiment

The second part of this book contains the evaluation of the data of the game playing experiment. We shall introduce some abbreviations and ways of speaking, which will then be used without further explanations.

A *gametype*, shortly written as (0,30), for example, represents the two-person bargaining game with incomplete information, where the initial random move selected the alternative value 0 for the first mover and the alternative value 30 for the second mover. The first component of the tuple always denotes the alternative of the first mover, and the second component denotes the alternative of the second mover. If the parameter value of α is indicated by this tuple representation, we sometimes refrain from explicitly mentioning it. Instead of (0,30), we might also write (L,H) for $\alpha = 30$. For each parameter value of α we distinguish four different gametypes.

The setup of the experiment was designed in such a way that each player played four times each of the four possible gametypes (H,H), (H,L), (L,H), and (L,L), but a gametype was experienced for the second time only if all other gametypes were experienced exactly once, and so on. Hence, we can distinguish four levels of experience for each gametype, the so called *experience levels* or, sometimes shortly called *levels*. We sometimes aggregate over the experience levels 1 and 2 and call the aggregation the *low experience level*, and aggregate over the experience levels 3 and 4 and call the aggregation the *high experience level*.

The term *outcome* should be understood as the non-discounted value a player received at the end of the bargaining. In case of agreement this is the share of the coalition value and it is the alternative value in case of conflict. The discounted value will be denoted as the *payoff*.

In the tables we shall sometimes abbreviate the term "average" by *avg* and the term "number" by *#*. For example, "#observations" stands for "number of observations".

The statistical tests which are used in the evaluation are described in detail in Appendix B. The application of a test to concrete problems is illustrated by a representative example. It is important that only whole sessions are counted as independent observations. Smaller units

of observations like individual plays or periods in plays cannot be assumed to be independent. The rejection of a null hypothesis based on such unjustified independence assumptions
would not support the alternative hypothesis, but rather a plausible lack of independence. The
way in which the tests are applied is always the same and will not be explained in detail in
every case, but only in Appendix B.

CHAPTER 6. THE AGREEMENT OUTCOMES

This chapter is dedicated to the analysis of the agreement outcomes in the game playing experiment. They will be taken as given, which means it will not be investigated how they emerged (this will be the task of the following chapters).

The analysis starts with the formulation of hypotheses on the bargaining outcomes that could be postulated before the experiment, based on the findings of previous experimental research. Afterwards, the agreement outcomes of the bargaining experiment are reported and the observations are summarized in résumés. They serve for the discussion of the hypotheses and the formulation of a descriptive theory.

6.1   HYPOTHESES ON THE AGREEMENT OUTCOMES

Although previous experiments were not concerned with the same bargaining situation, their findings can be used to hypothesize about the agreement outcomes to be expected.

In an experiment with incomplete information players face an unclear situation and previous experiments (see Section 4.1) suggest that players tend to approach the problem of non-transparency by applying familiar social norms, like fairness. By gaining experience, players may learn about the "bargaining power" of the different types and deviate to more competitive solutions.

The experiments by Hoffman and Spitzer (1982 and 1985) suggest that the players in our bargaining game should have a tendency to split the coalition value equally. One reason for this assumption is that the positions in our bargaining game are assigned by a random mechanisms and they are by no way earned by the players. In their terminology we have a no moral authority setup. Furthermore, our setup is not double blind, this means we did not ensure that the players were anonymous with respect to the experimenter. They were anonymous only with respect to the opponent. In this setup Hoffman et al. (1992) showed that subjects behave more egalitarian than under the double blind condition.

However, the experiment by Kuon and Uhlich (1993), which was conducted under the same experimental procedure, but with the exception that both players have complete information about the other player's alternative value, showed that subjects bargain competitively although they did not earn their positions and they are not anonymous to the experimenter.

Nevertheless, intrigued by the observations from previous experiments under incomplete information, we hypothesize that players will manage the unclear situation by applying familiar rules of fairness. It will be interesting to study whether this behavior changes over time. Hence, the question that remains is which social norms have to be applied in this bargaining situation.

To answer this question two possible cases will be distinguished. Firstly, both players may face the same alternative. If this is zero, fairness considerations (as well as normative theories of bargaining) propose that the players should, independent of the value of α, split the 100 equally. If both players face the high alternative they should also split the 100 equally if α is less or equal to 50. In the case where α is greater than 50, the game is not superadditive anymore and the strong players are better off by breaking off the negotiations and taking their alternative values. Secondly, the players may have different alternatives. Two different social fairness norms are applicable in this case: equality of absolute payoffs and equality of net payoffs (gains in addition to the alternatives). Henceforth, the two fairness concepts will be shortly called *equal split* and *equal split of the difference* (or *equal split of the surplus*).

Gamson (1961) introduced the *parity norm* as a fairness concept. It specifies that the rewards of the players should be distributed in direct proportion to their alternatives (contributions). Since in our model, where one player has the alternative value zero this concept is not applicable, no further investigations concerning the parity norm will be made.

However, the subjects in the underlying bargaining experiment have no knowledge about the other player's alternative and no instrument to force the other player to reveal his type. Therefore, each player has to choose between the application of the two fairness norms.

Hypothesis 1:   *For α < 50 the agreement outcomes will be in the range from the equal split up to the split of the difference with major peaks at the two fairness norms.*

Suppose α is greater than 50. A player of type H knows that it is only profitable to form a coalition with an opponent of type L. For this opponent the equal split of the 100 would be the most profitable fairness norm. However, this allocation is not reachable since an individually rational player of type H strives for at least one money unit in addition to his alternative value.

Hypothesis 2:   *If α > 50 a player of type H will only reach an agreement with a player of type L and his agreement outcome will be in the range between α + μ and α + ½(100−α).*

The variety of possible agreements for a player of type L is very large in the case of α > 50. The range of agreement outcomes is determined through three possible situations. The opponent of a weak player may also be weak and they may identify each other and agree on the equal split. It may also be possible that the opponent is weak and pretends to be strong by insisting on a high payoff like the split of the difference in addition to the high alternative value. Since a player of type L has no possibility to verify whether the other player is also weak or strong, he may want to accept this offer, otherwise fearing the risk of break off. This may lead to the outcome of the split of the difference, which determines the lower bound of the range. The weak player himself may also try to pretend that he is strong and by this have the chance to receive the split of the difference in addition to the high alternative value, which is the "best" outcome for him and determines the upper bound of the range.

Remember that Hoggatt et al. (1978) observed equal splits only in one quarter of all plays of two weak players.

Hypothesis 3:   *If α > 50, the agreement outcome of the weak player is in the range between the equal split of the difference ½(100−α) and the equal split of the difference in addition to the high alternative value ½(100+α).*

All hypotheses specify a range of possible agreement outcomes. From previous experiments (inter alia Uhlich, 1990, and Kuon et al., 1993) it is known that the outcomes experimental subjects reach are not concentrated on a single point. Due to different bargaining skills a whole area of agreement outcomes may be reached. However, not all points in this area are reached equally often. Experimental subjects have a tendency to choose "round" numbers so that these will occur more likely than other possible outcomes.

Hypothesis 4:   *The agreement outcomes will not be distributed equally in the assumed ranges, but will have peaks on the prominent numbers.*

In previous experiments (Kuon and Uhlich, 1993) it happened very rarely, and was viewed as a typing error, that a subject agreed on an allocation which yielded a payoff less than the alternative value. In the present experiments we also expect not to observe non-individually rational (perhaps altruistic) behavior. All plays of two high type players in games with $\alpha > 50$ should end in conflict.

Since for $\alpha < 50$ all possible games are superadditive and this is common knowledge, all plays could end with an agreement where both players exceed their alternatives. From previous experiments (inter alia Uhlich, 1990 and Kuon et al., 1993) it is known that plays ended with a break off although the underlying game was superadditive. This fact can be explained by punishment of the proposal of an unfair allocation (see the discussion of Kahneman, 1986a in Section 4.2) or a too large difference in the relative payoffs (see the discussion of Bolton, 1991 in Section 4.2).

Hence, we also must expect to see this phenomenon and may hypothesize an increasing number of break offs for a smaller surplus. A small surplus induces a small number of possible agreements with small gains for the players. In this situation a punishment is "less costly" for the punisher and therefore seems to be more likely.

Hypothesis 5:   *For $\alpha > 50$ two high type players will never agree. In all other cases the number of break offs will be in negative correlation to the size of the surplus to divide.*

6.2   THE AGREEMENT OUTCOMES

The experiment examined five different values for the parameter α: 30, 45, 55, 60, and 70. For each of these parameter values we distinguish three different types of games, namely that both players have the alternative 0, or both players have the alternative α, or finally that one player has the high and the opponent has the low alternative. For the first and the second type of game 72 plays were played in the experiment; 12 in each of the 6 independent subject groups. For the case where both players have different alternative values we have 144 observation; 72 from the case that the low type player is moving first and 72 from the case that the high type player is the first mover.

Since, for every parameter value of α, the Wilcoxon matched-pairs signed-ranks test did not reject the hypothesis that there is no difference in the outcome of a first mover and the outcome of a second mover, we shall neglect the distinction of the first mover and the second mover here.

The observations from the experimental data will be summarized by résumés.

THE GAMES WITH $\alpha < 50$

Table 6.1 shows the distribution of the agreement outcomes for the plays of games with $\alpha=30$ and $\alpha=45$. For the case that both players have the same alternative the table shows the higher of the two agreement outcomes. In the other case the outcome of the player with the high alternative is given. (Note that only the numbers with an occupation of at least 1 are listed).

Table 6.1: Distribution of the agreement outcomes in the plays of games with $\alpha < 50$

Number of agreement outcomes on...	$\alpha = 30$			$\alpha = 45$		
	(0,0)	(30,0)	(30,30)	(0,0)	(45,0)	(45,45)
48		1			1	
49		5			1	
50	55	108	42	55	81	21
51	6	3	6	4	5	3
52				5	6	1
53		1	1	1	1	
54	1		1	1	1	
55	2	3		2	6	
56					1	
57						
58	1					
59		1				
60	2	1		2	8	
65	2	1				
66		1				
68		1				
69					1	
70	1			1		
71					1	
85	1			1		

The agreement outcomes of the games with $\alpha < 50$ look very similar to each other. More than three quarters are on the equal split of the coalition value. The remaining outcomes are spread between 50 and 85, but mostly with one single observation at each point. There is no other focal point besides 50 that seems to have an importance. The largest occupation among all other points can be found on 51.

With the level of experience in the bargaining game, the tendency towards the equal split agreement rises. Table 6.2 shows the average deviation from the equal split (for the player with the higher outcome) in the different experience levels of the experiment. Since for all four gametypes of each alternative value these numbers are similar, for the sake of simplicity, only the average over all gametypes is given.

Table 6.2: Average deviation from the equal split in the different experience levels

	Experience level			
	1	2	3	4
$\alpha = 30$	2.41	.87	.42	.18
$\alpha = 45$	2.75	1.07	.58	.69

The table shows a clear convergence toward the equal split for more experienced players.

Table 6.3 provides the information about the proportion of equal split agreements (among all agreements). Moreover, it presents the agreement rate in the plays of the games with $\alpha < 50$.

Table 6.3: Agreement and conflict in plays of games with $\alpha < 50$

	$\alpha = 30$			$\alpha = 45$		
	(0,0)	(30,0)	(30,30)	(0,0)	(45,0)	(45,45)
Number of conflicts	1	18	22	0	31	47
Number of agreements	71	126	50	72	113	25
Proportion of agreements	.986	.875	.694	1	.785	.347
Prop. of 50:50 outcomes among all agreements	.775	.857	.840	.763	.717	.840

Résumé 1:   *For the plays of games with $\alpha = 30$ and $\alpha = 45$, independent of the alternatives of the players, more than three quarters of the agreement outcomes coincide with the equal split of the coalition value. For experienced players the average deviation from the equal split agreement is less than 1.*

The order test was applied to examine whether the proportion of agreements falls with the shrinkage of the surplus. For $\alpha = 30$ and $\alpha = 45$ the one-sided order test rejects the null hypothesis that there is no trend in the agreement rates in favor of the alternative of a decreasing trend in the agreement rates for a shrinking surplus at a significance level of .001. This means that the gametype (0,0) has the highest agreement rate, followed by the game-types (0,α) and (α,0), and that (α,α) has the lowest agreement rate.

Résumé 2:   *The proportion of agreements falls with the shrinkage of the surplus. This means it has the highest value for the case that both players have the alternative 0 and the lowest value for the case that both players have the alternative α.*

THE GAMES WITH $\alpha > 50$

As expected, in cases where both players have the high alternative (almost) always a conflict was reached. In one session with plays of games with $\alpha = 60$ the same player twice accepted an offer that yielded himself less than his outside option.

Résumé 3:   *When both players have an alternative value that is greater than 50 (almost) no agreements were reached.*

Table 6.4: Agreement and conflict in plays of games with $\alpha > 50$

	$\alpha = 55$		$\alpha = 60$		$\alpha = 70$	
	(0,0)	(55,0)	(0,0)	(60,0)	(0,0)	(70,0)
Number of conflicts	3	78	4	98	3	118
Number of agreements	69	66	68	46	69	26
Proportion of agreements	.958	.458	.944	.319	.958	.181

For all parameter values of $\alpha > 50$ the one-sided order test rejects the null hypothesis that there is no trend in the agreement rates in favor of the alternative of a decreasing trend in the agreement rates for a shrinking surplus at a significance level of .001.

Résumé 4:   *If both players have the alternative 0 the proportion of agreements is close to 1. In the case that one player has the high alternative, the proportion of agreements is lower than ½ and shrinks with increasing α, for $\alpha > 50$.*

The distribution of the agreement outcomes for $\alpha > 50$ will be illustrated in the figures 6.1 to 6.6. In plays of games where both players have the same alternative, the higher agreement value is shown in the figures and otherwise the outcome of the player with the high alternative. The figures provide a level indicator for every outcome. The first two levels (*low experience level*) and the last two levels (*high experience level*) are aggregated to simplify the readability of the figures. The observations are summarized in the following résumés.

Résumé 5:   *The agreement outcome of a low type player playing a low type player varies in the range from 25 to 75 with clear peaks on the numbers divisible by 5, for $\alpha = 55$.*

Résumé 6:   *The agreement outcome of a high type player playing a low type opponent varies in the range from 50 to 82 with clear peaks on the numbers divisible by 5, for $\alpha = 55$.*

Résumé 7:   *The agreement outcome of a low type player playing a low type opponent varies in the range of 20 to 80 with clear peaks on the numbers divisible by 5, for $\alpha = 60$.*

Résumé 8:   *The agreement outcome of a high type player playing a low type opponent varies in the range from 61 to 91 with clear peaks on the numbers divisible by 5, for $\alpha = 60$. Two non-individually rational outcomes on 21 and 30 are observed.*

Résumé 9:   *The agreement outcome of a low type player playing a low type opponent varies in the range of 15 to 85 with clear peaks on the numbers divisible by 5, for $\alpha = 70$.*

Résumé 10:   *The agreement outcome of a high type player playing a low type opponent varies in the range from 50 to 90 with clear peaks on the numbers divisible by 5, for $\alpha = 70$.*

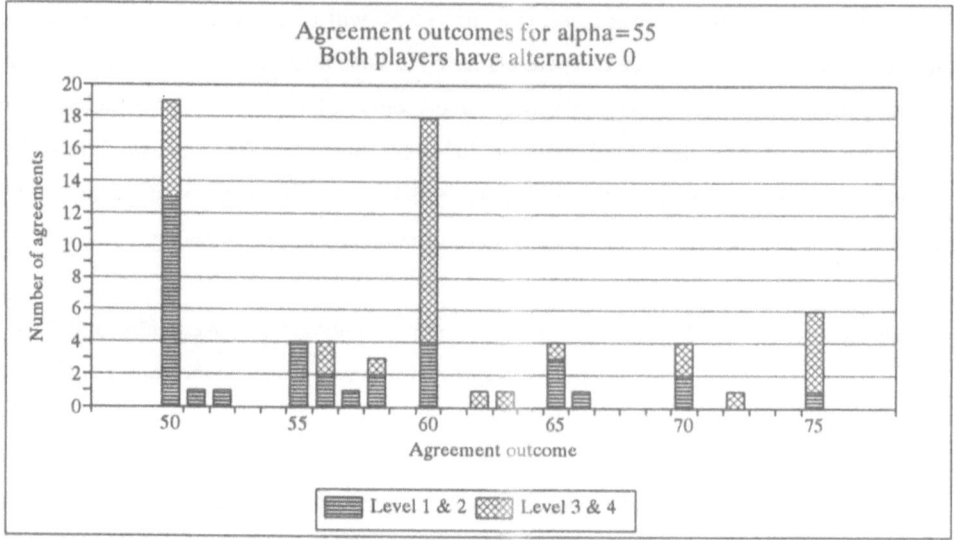

Figure 6.1:   Distribution of agreement outcomes of two weak players for $\alpha = 55$

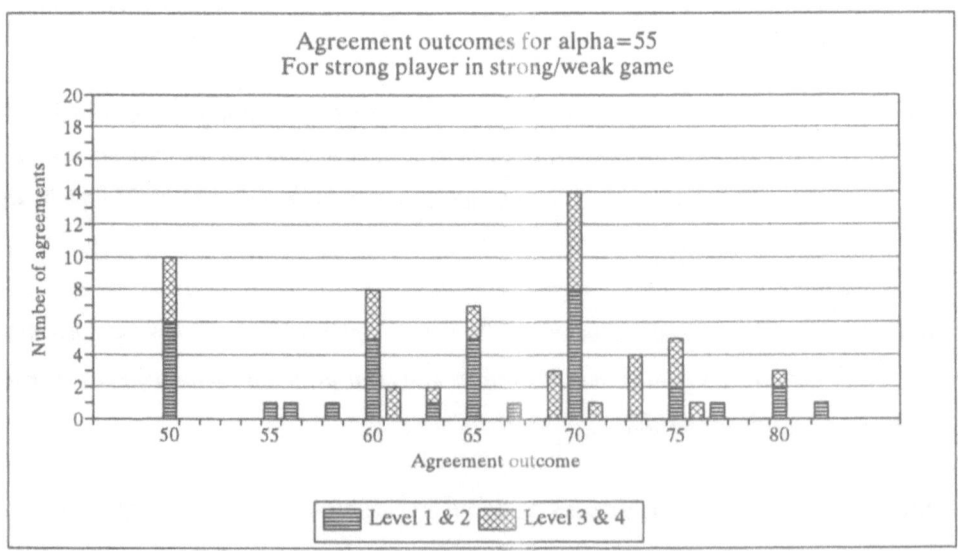

Figure 6.2:   Distribution of agreement outcomes of a strong player playing a weak player for $\alpha = 55$

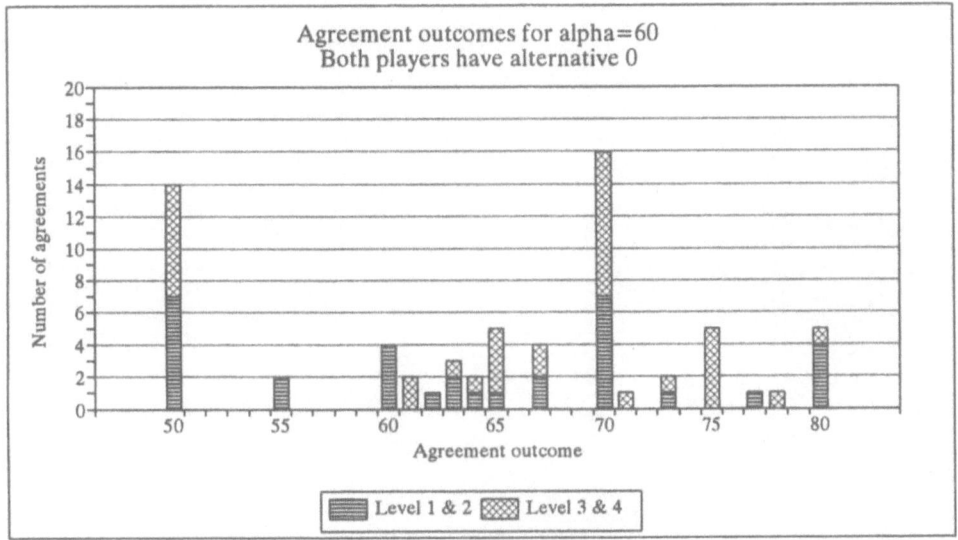

Figure 6.3:   Distribution of agreement outcomes of two weak players for $\alpha=60$

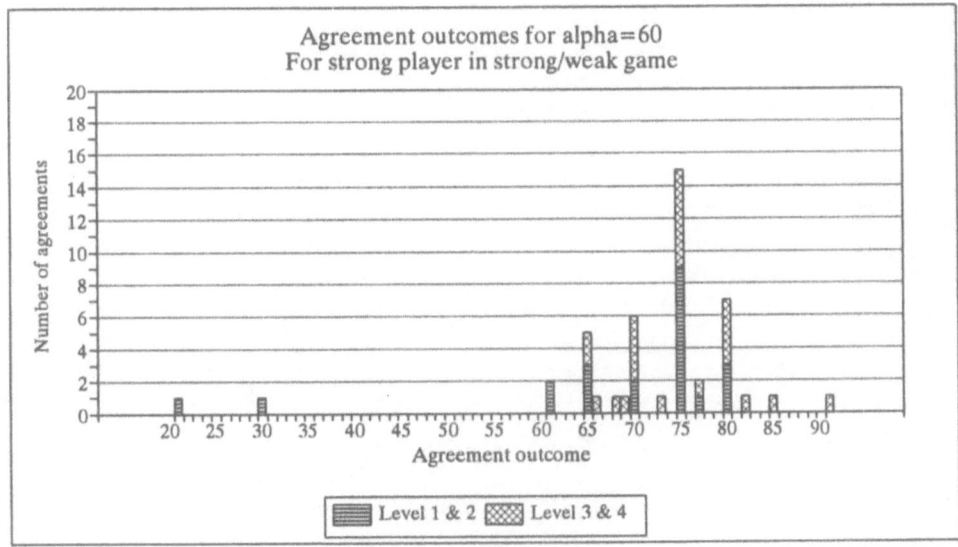

Figure 6.4:   Distribution of agreement outcomes of a strong player playing a weak player for $\alpha=60$

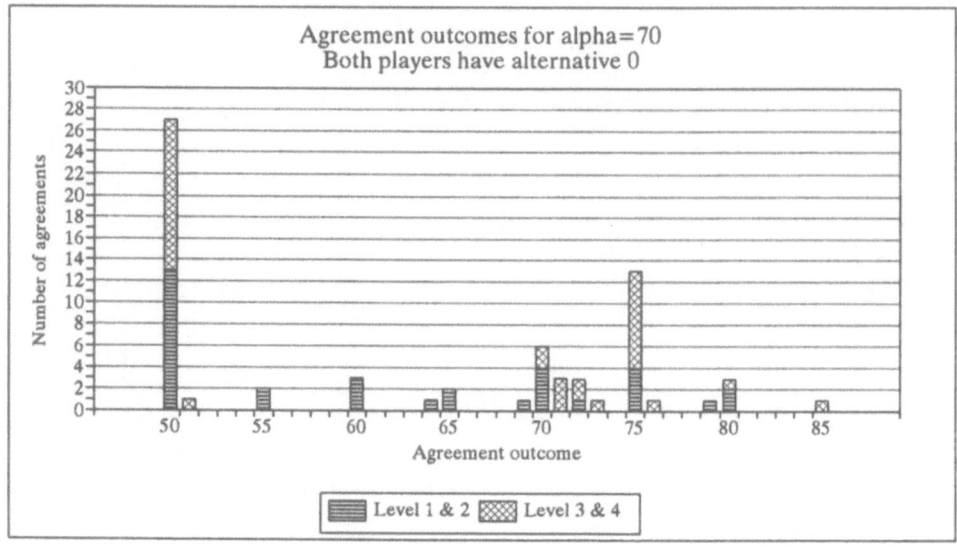

Figure 6.5:   Distribution of agreement outcomes of two weak players for $\alpha = 70$

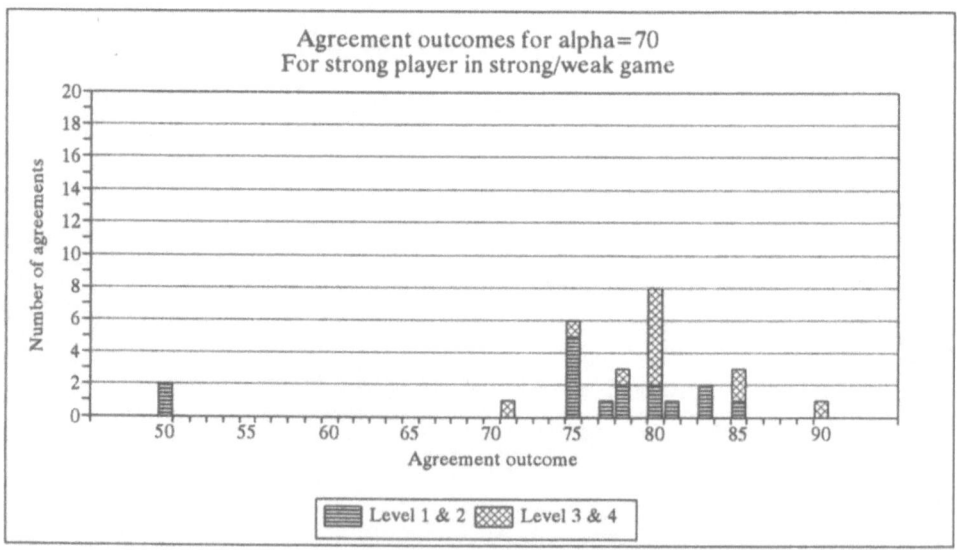

Figure 6.6:   Distribution of agreement outcomes of a strong player playing a weak player for $\alpha = 70$

There is one unexpected fact in the experimental data. Type H players agree to the equal split in plays of a weak and a strong player for $\alpha=55$. This happens for unexperienced as well as for experienced players. Since this occurs in 10 out of 66 plays and can be found in five of the six subject groups, this cannot be seen just as errors made by some players. These agreements may be caused by altruism. For a discussion of the phenomenon of altruism see Section 8.5. Furthermore, we observe an agreement outcome of 21 and an agreement outcome of 30 for a strong player in the first experience level for $\alpha=60$ and two agreement outcomes of 50 for the strong player for $\alpha=70$ (also in the low experience phase). Since these are only few observations in the low experience level, the author tends to view them as "agreements by mistake" and refrains from a further consideration.

It is a very remarkable fact that the majority of agreement outcomes in plays of two weak players in the case of $\alpha=70$ is on the equal split. This might be explained by the great risk of break off a player faces in such a situation. If a player pretends to be strong, the other one might break off immediately. So, revealing the true type (by proposing the equal split) seems to be a common behavior for players of type L.

This, however, does not explain why the weak opponent accepts the proposal of the equal split. He might also try to reach a high outcome by imitating the strong player. It cannot be the fear of a break off by the weak player playing a strong player. Table 7.7 in Section 7.2 shows that break offs by the weak player in plays with a strong player occur rather seldom. However, if the weak player breaks off the major reason is that he received a maximal offer of less than 50 (table 7.8 in Section 7.2). On the other hand table 7.10 in Section 7.3 and the following explanations show that in the majority of the cases the weak player insisted on the equal split. This means that the weak player revealed by the demand of 50 and repeated it continuously. In consequence strong opponents break off (see table 7.10), while weak opponents tend to accept this proposal. The reasoning of the weak revealer might be that this behavior yields an expected payoff of 25 (having an equal split agreement in half of the cases and a conflict in half of the cases) and, on the other hand, in an agreement with the strong opponent (or a bluffing weak opponent) he might expect less than 25.

Changes by experience

Does the distribution of the agreement outcomes change over the experience levels? To answer this question the order test is applied to the following values: for each play the absolute value of the difference between the agreement outcomes of the players is calculated, and then the average over all plays of an independent subject group is evaluated. Since the sum of two agreement outcomes is 100 in each play (and therefore never changes), we look at the absolute value of the difference of both agreement outcomes. Except for $\alpha = 60$ and $\alpha = 70$ the one-sided order test rejects the null hypothesis of the non-existence of a trend. For the parameter values of $\alpha < 50$ the alternative of a decreasing trend was favored (at a significance level of .01 and .03, respectively, one-sided). For $\alpha = 55$ the alternative of an increasing trend is favored (at a significance level of .01, one-sided). For $\alpha = 60$ and $\alpha = 70$ the two-sided order test cannot reject H_0 at a significance level of .2.

A decrease in the average of the absolute value of the difference between both agreement outcomes means that the two agreement outcomes become more equal with higher experience. This means they converge to the equal split. We already found this result in table 6.2. While in the first experience level the average deviation from the equal split was about 2.5, it was clearly below 1 in the fourth experience level. For $\alpha = 55$ the average difference increased with experience. This means that the asymmetry in the agreement outcomes increases with experience.

The analysis shows that one has to distinguish between the inexperienced and the experienced behavior of subjects. If subjects face a new situation it is approached with familiar norms that are transferred to this situation. But, as the subjects gain experience with the situation these familiar norms are displaced by the experience gained in the situation. In the two-person bargaining game the weak types experience their bargaining power in the plays of games with $\alpha > 50$. They find out that they can reach high outcomes. This experience displaces the inexperienced behavior of more egalitarian outcomes. A very similar observation was made in a three-person quota game experiment by Selten and Kuon (1993). Inexperienced players are guided by equal splits of the coalition values, while with experience quotas gain more influence on behavior. On the other hand, experience can also lead to a confirmation of the familiar social norms, like this was observed for the egalitarian distribution in plays of games with $\alpha < 50$.

6.3   THE PROMINENCE LEVEL

The idea of prominence was first introduced by Schelling (1960). Investigations of the prominence in the decimal system are due to Albers and Albers (1983), Tietz (1984) and Selten (1987). They suggest that numbers are perceived as "round", if they are divisible without remainder by a prominence level, which depends on the data context. The *prominence level* is of the form $\Delta = \gamma 10^{\eta}\mu$, with $\gamma = 1, 2, 2.5, 5$, $\eta = 0, 1, 2, ...$, and μ the smallest money unit. We use the method proposed by Selten (1987) to determine the prominence level for each independent subject group and obtain the prominence level of 5 in each case. The agreement outcomes already indicated that the subject are guided by "round" numbers, here especially numbers that are divisible by 5.

Résumé 12:   *The prominence level in each session is 5.*

Figure 6.7 displays the percentage of *prominent numbers* (numbers divisible by 5 without remainder) in the demands for the players' own outcome. The vast majority of more than 70% of all demands is on a prominent number.

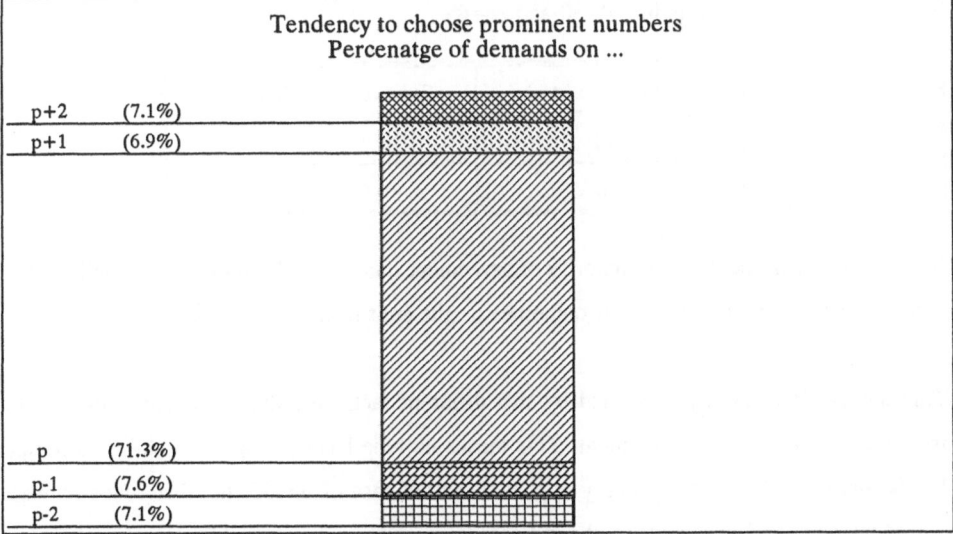

Figure 6.7:   Tendency to choose prominent numbers. p stands for a prominent number (a number divisible by 5 without remainder).

6.4   VERIFICATION OF THE HYPOTHESES

Section 6.1 presented the hypotheses on the agreement outcomes of the bargaining experiment that were stated in the light of previous experiments on games with incomplete information and experiments on fairness behavior. In this section we shall discuss the validity of the hypotheses with respect to the actual experiment. As a reminder the hypotheses will be repeated before discussing them.

Hypothesis 1:   *For $\alpha < 50$ the agreement outcomes will be in the range from the equal split up to the split of the difference with major peaks at the two fairness norms.*

This hypothesis failed in the light of the experimental data. Its formulation was based on the assumption that a player may choose between the application of two fairness norms, the equal split and the split of the difference. Since no conformity about a unique fairness norm could be expected, it was assumed that the agreement outcomes fall into the range that is spread by the two norms. For the two parameters of $\alpha < 50$ the table 6.5 shows the value of each fairness norm.

Table 6.5: Fairness norms for $\alpha = 30$ and $\alpha = 45$

	Equal split	Split of the difference
$\alpha = 30$	50	35
$\alpha = 45$	50	27.5

In the experiment the striking majority of agreement outcomes was on the equal split. There is no tendency towards agreement outcomes on the split of the difference.

This observation may by explained by two different facts. Firstly, each type improves his payoff in case of a 50:50 agreement. This means no type has the "right" to claim more than 50. Secondly, and this is especially applicable to the case of $\alpha = 30$, the difference between the equal split and the split of the difference is relatively small. Since the payoffs are discounted with each new proposal, subjects seem to like a fast agreement on the "prominent" outcome of the equal split rather than a long bargaining over a few points.

Conclusion 1:   *In plays of games for $\alpha < 50$ the striking majority of agreement outcomes is on the equal split, independently of the type of the player. On a higher level of experience the tendency towards the equal split becomes even clearer.*

Hypothesis 2:   *If $\alpha > 50$ a player of type H will only reach an agreement with a player of type L and his outcome will be in the range of $\alpha + \mu$ and $\alpha + \frac{1}{2}(100 - \alpha)$.*

For $\alpha = 55$ and $\alpha = 70$ no agreements occurred in plays of two strong players. For $\alpha = 60$ we observed two agreements in plays of two strong players. In both cases it was the same player who agreed on an allocation that yielded himself less than his alternative. By the way, it was the same player who accepted a proposal of 30 as a strong player in a play with a weak player (see figure 6.4). Remember, that we observed agreements yielding non-individually rational outcomes for the strong player in games with $\alpha = 55$ (figure 6.2) and $\alpha = 70$ (figure 6.6). But all these agreements occurred in plays with the weak player.

Table 6.6 gives for each of the three values of α the range between $\alpha + \mu$ and the split of the difference in addition to α, in order to examine whether the hypothesized range is of descriptive validity.

Table 6.6: $\alpha + \mu$ and the split of the difference in addition to α for $\alpha = 55$, 60, and 70

	$\alpha + \mu$	$\alpha + \frac{1}{2}(100 - \alpha)$
$\alpha = 55$	56	77.5
$\alpha = 60$	61	80
$\alpha = 70$	71	85

In view of the smallest money unit of 1, an agreement of 77.5 was not reachable by the subjects and it would have to be rounded to 77 or 78. But, with respect to the tendency of the subjects to choose prominent numbers we shall round this value to the next prominent number, which is 80 (at the prominence level of 5). The respective values for $\alpha = 60$ and $\alpha = 70$ are already prominent numbers.

Comparing these ranges with the actual outcome ranges in the experiment (see résumés 6, 8, and 10), the upper bounds seem to fit the experimental data quite well. For $\alpha=55$ we observe one agreement outcome of a strong player in a play with a weak player above 80 (on 82, see figure 6.2), for $\alpha=60$ we observe three agreement outcome above 80 (on 82, 85, and 91, see figure 6.4), and for $\alpha=70$ one agreement outcome of a strong player in a play with a weak player is located above 85 (on 90, see figure 6.6).

But, the experiment also shows that the vast majority of players is not satisfied only with one smallest money unit in addition to the alternative value. Instead they strive for at least one prominence level above the alternative value. From a behavioral viewpoint this is not surprising. Experiments on ultimatum bargaining (see Section 4.2) showed that subjects are not satisfied with the smallest money unit. They want to win a "considerable amount of money". The "smallest acceptable gain" is a very personal quantity, but in our experiment a good approximation seems to be the prominence level.

This leads to the following payoff bounds for the strong player:

Table 6.7: Bounds for the agreement outcome of the strong player for $\alpha=55$, 60, and 70

	Agreement bounds for the strong player	
	lower bound	upper bound
$\alpha = 55$	60	80
$\alpha = 60$	65	80
$\alpha = 70$	75	85

The predictive power of these bounds with respect to the experimental data will be studied in Section 6.6. Therefore, at this point, no conclusion will be formulated concerning hypothesis 2.

Hypothesis 3:   *If $\alpha > 50$, the agreement outcome of the weak player is in the range between the equal split of the difference $\frac{1}{2}(100-\alpha)$ and the equal split of the difference in addition to the high alternative value $\frac{1}{2}(100+\alpha)$.*

Like in the experiment of Hoggatt et al. (1978), the majority of the agreement outcomes in plays of two weak players deviates from the equal split. Nevertheless, the question arises whether they really scatter in the whole range from the split of the difference to the split of the difference in addition to α. The answer (given by the data of the experiment) is: nearly. Remember, that the agreement outcomes are shown in figures 6.1 to 6.6.

For $\alpha=60$ and $\alpha=70$, where the bounds are on prominent numbers, they are confirmed by the experiment. For $\alpha=55$ the calculation of the equal split of the surplus yields the non-achievable number of 22.5. Analogously, to the case of the bound of the strong player, the equal split of the surplus in addition to α will be rounded to the next prominent number. The experimental data suggest that the interval should not be enlarged by the adjacent prominent numbers outside the interval, but should be set to [25,75].

In games of a weak and a strong player we already concluded that the strong player seems to strive at least for a prominence level in addition to his alternative value. This means that the weak player receives at least the split of the difference and at most he can expect to get the surplus minus the prominence level. For $\alpha=60$ and $\alpha=70$ these intervals fall into the outcome range of two weak players. For $\alpha=55$ this interval ranges from 20 to 40. This means that the lower bound has to be corrected to 20 in this case.

These considerations lead to the bounds given in table 6.8.

Table 6.8: Bounds for the agreement outcome of the weak player for $\alpha=55$, 60, and 70

	Agreement bounds for the weak player	
	lower bound	upper bound
$\alpha = 55$	20	75
$\alpha = 60$	20	80
$\alpha = 70$	15	85

This opens a wide range for the agreement outcomes of the players of type L. The delineation of the agreement range is a weak "everything is possible" statement, but there seems to be no reasonable way to strengthen this result. The predictive power of these bounds with respect to the experimental data will be studied in Section 6.6. Therefore, at this point, no conclusion will be formulated concerning hypothesis 3.

<u>Hypothesis 4:</u>   *The agreement outcomes will not be distributed equally in the assumed ranges, but will have peaks on the prominent numbers.*

This hypothesis concerns the distribution of the agreements over the proposed ranges. The figures 6.1 to 6.6 in Section 6.2 show very strikingly that the distribution is not uniform. The prominent numbers (numbers divisible by 5) are chosen with a high frequency. The ranges of the agreements outcomes may be very large, but the subjects mostly choose prominent numbers.

<u>Conclusion 4:</u>   *In the agreement ranges the prominent numbers are strikingly more often chosen than the other numbers.*

<u>Hypothesis 5:</u>   *For $\alpha > 50$ two high type players will never agree. Otherwise, the number of break offs will be in negative correlation to the size of the surplus to divide.*

For $\alpha = 60$ we observed two agreements of two high types players. For all parameter values of α the one-sided order test showed the result stated in conclusion 5 (see Section 6.2).

<u>Conclusion 5:</u>   *The proportion of conflicts increases with the shrinkage of the surplus.*

6.5 ASYMMETRY IN THE AGREEMENT OUTCOMES OF TWO WEAK PLAYERS

In the case of a bargaining of two weak players and $\alpha > 50$ many subjects could achieve a higher payoff than the equal split. This section shall study several reasons that might lead to asymmetric agreements, independently of the player's bargaining skill. We shall start with posing the question whether there is a first mover advantage (disadvantage) in these games which "automatically" yields a higher (lower) agreement outcome for the first mover. Secondly, we shall investigate whether the diversity in outcomes is caused or diminishes by the experience of the subjects. Thirdly, we shall look at the average length of a bargaining game. The question arises whether a relatively longer bargaining is necessary to result in highly asymmetric agreement outcomes. If this would be true, the discount factor could make this outcome less profitable than a more egalitarian one.

FIRST MOVER ADVANTAGE/DISADVANTAGE

To study whether the first moving player is a priori in a better (worse) position concerning his outcome, the Wilcoxon matched-pairs signed-ranks test was conducted for each parameter constellation of α. Especially for the plays of games of two low type players it was examined whether there is a difference in the outcome of a first and a second mover. The null hypothesis that there is no difference could not be rejected for each parameter value of α.

THE AGREEMENT OUTCOMES AT DIFFERENT LEVELS OF EXPERIENCE

In Section 6.2 we already found that the asymmetry in the agreement outcomes of the two players rises with experience for $\alpha = 55$. This result was deduced without a distinction of the types of the players. If we only look at plays of two weak players and apply the one-sided order test to the average of the absolute value of the difference between both agreement outcomes a similar result can be found. For $\alpha = 55$ the alternative of a increasing trend is favored at a significance level of .01 (one-sided). For $\alpha = 60$ and $\alpha = 70$ the null hypothesis of the non-existence of a trend cannot be rejected at a level of .2 (two-sided).

AVERAGE LENGTH OF BARGAINING

How long is the average bargaining game that leads to a certain payoff? Does it take too much effort (too many bargaining steps) to achieve a highly asymmetric result? Table 6.9 aims to answer these questions. The table contains only the agreement outcomes on prominent numbers since only for these numbers there are enough observations to meaningfully calculate an average.

Table 6.9: Average length of bargaining for agreement outcomes on prominent numbers

Agreement outcome	Average length of negotiations		
	$\alpha = 55$	$\alpha = 60$	$\alpha = 70$
50	3.68	10.92 [2]	10.03
55	3.50	7.50	2.50
60	2.94 [1]	4.25	6.30
65	3.00	8.40	10.50
70	4.50	11.47	2.50
75	6.67	6.17	3.62
80	-	3.80	2.00
Conflicts	14.33	29.00	21.30
[1] Outlier 18 taken out of a sequence of 18 numbers from 1 to 8 (otherwise avg 8.5)			
[2] Outlier 58 taken out of a sequence of 14 numbers from 1 to 35 (otherwise avg 14.29)			

In the case of $\alpha=55$ the bargaining lengths for achieving the outcomes between 50 and 70 are not significantly different. The median of 3.5 shows that they are quite short. An exception is the higher effort of an average bargaining length of 6.67 that was needed to reach the outcome of 75. But, the seven times discounted value of 75 yields an actual payoff of 69.90 which is higher than a four times discounted value of 50, which is 48.03. This shows that the longer bargaining time does not destroy the advantage of the higher outcome. Indeed, with the discounting taken into account the outcome of 75 yielded the highest final payoff.

In the plays of games with $\alpha=60$ the majority of agreements was on the values 50 and 70, which both have an average bargaining length of approximately 11. The overall greater number of steps than in $\alpha = 55$ indicates a greater diversity in the subjects' opinion about the bargaining outcome.

Interestingly, the equal split agreement was reached after an average of 10 steps in the case of $\alpha = 70$. This indicates that either the revelation of one player was delayed or that the other player tried to exploit the revealer. However, the second most frequent outcome of 75 was reached after only 4 steps.

It can be concluded that if a highly asymmetric agreement is reached, the effort in terms of discount steps is so small that this outcome is the most profitable one. Clearly, one has to pose the question whether the effort in terms of conflicts is not too high for asymmetric agreements. This question will be raised in Chapter 7. Now we are only interested in the realized agreements.

Table 6.9 gives as an additional information the average number of steps before a conflict was reached. For all parameters of α this number is at least twice as high as the longest agreement length, which indicates that weak players are very patient in bargaining before breaking off. This is not surprising since they always improve their payoff by accepting. Nevertheless, non-monetary motivation leads to break offs.

6.6   DESCRIPTIVE THEORIES OF THE AGREEMENT OUTCOMES

The prediction of the agreement outcomes in a game of incomplete information is a difficult task. Only in rare cases normative theories exist, and they mostly predict a wide range of possible agreements such that they give no clear picture of the expected bargaining outcomes. Descriptive theories for games with incomplete information do not exist.

For the two-person bargaining game which follows the same experimental procedure as the game considered here, but where the players have complete information about the alternative values of both players, a descriptive theory was proposed by Uhlich (1990) and further tested and extended by Kuon and Uhlich (1993). This theory is called the *Negotiation Agreement Area (NAA)*. The NAA predicts an area for the agreement outcomes, dependent on the characteristic function of the game. The basic idea is that the players' initial demands are close to the maximal achievable outcome and that the bargaining proceeds with relatively equal concessions from there on. To calculate the maximal achievable outcome a player has to be aware of the alternative value of the opponent. Since, this crucial condition is violated in the game with incomplete information one should not expect the predictive power of the NAA to be as excellent as in the games of complete in formation. Nevertheless, in absence of other descriptive theories we shall apply the NAA to the game of incomplete information and test its predictive power.

Moreover, we shall look at two other distribution schemes. The verification of the hypotheses on the agreement outcomes in Section 6.4 showed that the "fair" solutions described by the equal split for $\alpha < 50$ and the split of the difference for $\alpha > 50$ may be successful predictors for the borders of the range of the agreement outcomes. A very intriguing fact was that the outcomes were not equally distributed over this interval, but had major peaks on the prominent numbers. Therefore, one can look at the revision of the "fair" solution concept by the concentration on the prominent numbers in the "fair" solution area. Moreover, it is interesting to look at the predictive power of the prominent numbers in general.

In what follows the descriptive concepts to be tested are described in more detail. The variable u_1 denotes the lower outcome bound for the strong player and u_2 denotes the lower outcome bound for the weak player.

THE NEGOTIATION AGREEMENT AREA (NAA)

The *Negotiation Agreement Area* (NAA) (see Uhlich 1990, and Kuon and Uhlich 1993) is a descriptive area theory for two-person bargaining games in characteristic function form with complete information. The NAA assumes that the final agreement is reached by relatively equal concessions, starting from the values which are perceived as maximal attainable by the players. In case that both players have the same alternative value, the NAA predicts an equal split of the coalition value. In case of different alternative values, the strong player views the whole surplus as maximally attainable, while the weak player views a value between the equal split of the surplus in addition to his alternative value and the whole surplus as the maximal attainable value.

The application to the characteristic function v of the present experiment ($v(12)=100$ and $v(i)=$alternative value of player i) yields that in case of different alternative values, the maximal attainable value for the strong player (A_1^{max}) is the coalition value 100. For the weak player it is between the equal split of the surplus (A_2^{min}) and $100-\alpha$ (A_2^{max}). Considering the assumption of relatively equal concessions, the lower bounds b_1 and b_2 for the strong and the weak player's outcome, respectively, are given by:

$$b_1 := \frac{A_1^{max}}{A_1^{max} + A_2^{max}} v(12) \text{ and } b_2 := \frac{A_2^{min}}{A_1^{max} + A_2^{min}} v(12),$$

which is equivalent to:

$$b_1 := \frac{100}{200-\alpha} 100 \text{ and } b_2 := \frac{100-\alpha}{200-\alpha} 50.$$

With respect to the prominence in the data, these bounds are adjusted to the lower adjacent prominence level but at least to $v(i)+\mu$. The adjusted lower bounds u_i are given by

$$u_i := max[v(i)+\mu, \Delta int(b_i/\Delta)], \quad i=1, 2.$$

Recall, that Δ denotes the prominence level of the data set, and μ denotes the smallest money unit. In the present experiment they have the values $\Delta=5$ and $\mu=1$. The value int(x) denotes the highest integer value not exceeding $x \in \mathbf{R}$.

The *Negotiation Agreement Area* (NAA) predicts the set of all grid points (x_1, x_2) with the properties $x_1 + x_2 = 100$ and $x_i \geq u_i$, $i = 1$, 2. If $v(1) = v(2)$, the NAA predicts an equal division of the coalition value, i.e. $x_i = 50$.

The predicted lower bounds of the NAA in case that both players have different alternative values are given in table 6.10.

Table 6.10:   Prediction of the NAA, if both players have different alternative values

α	b_1	u_1	b_2	u_2
30	58.82	55	20.59	20
45	64.52	60	17.74	15
55	68.97	65	15.52	15
60	71.43	70	14.29	10
70	76.92	75	11.54	10

THE FAIR DISTRIBUTION SCHEME

The *fair distribution scheme* (FDS) combines the two distribution schemes based on the fairness norms equal split and equal split of the difference. For $\alpha < 50$ it predicts the equal division of the coalition value, and for $\alpha > 50$ it predicts that the strong player receives at least a prominence level in addition to his alternative value, and that the weak player receives at least the equal split of the coalition value adjusted to the adjacent lower prominent number. Notice, that we already introduced a modification of the "pure" distribution scheme by fairness norms, by lowering the bound for the strong player. The experiments showed that the strong player is not able to always enforce the split of the surplus in addition to α as an agreement outcome. Since he has no possibility to "prove" that he is actually strong, he often remains only with a prominence level in addition to α. The above formulation already considers this observation. Therefore, the FDS combines this ex post insight with the fairness norms. Since it is obvious that a purely fairness based distribution scheme would perform worse it is not considered here.

Accordingly, the *fair distribution scheme* (FDS) predicts the set of all grid points (x_1, x_2) with the properties $x_1 + x_2 = 100$ and $x_i \geq u_i$, $i = 1, 2$, for

$u_1 = 50$ and $u_2 = 50$, for $\alpha < 50$,

$u_1 = \alpha + \Delta$ and $u_2 = \Delta \mathrm{int}(\frac{1}{2}(100 - \alpha)/\Delta)$, for $\alpha > 50$, where $\Delta = 5$ is the prominence level.

THE PROMINENT FAIR DISTRIBUTION SCHEME

The *prominent fair distribution scheme* (PFS) combines the FDS distribution scheme with the insights on the use of prominent numbers in experiments. The intervals predicted by the FDS are very large, and the distribution of the agreement outcomes is highly unequal. Therefore, the combination of the fair solution concept and the prominence concept only predicts the prominent numbers in this interval. Hence, the set of the predicted outcomes is no longer a connected interval, like in the NAA and the FDS, but a set of prominent numbers, which have the common property that they all fall into the interval of the FDS.

Accordingly, the *prominent fair distribution scheme* (PFS) predicts the set of all grid points (x_1, x_2) with the properties $x_1 + x_2 = 100$, $x_i \geq u_i$, $i = 1, 2$, and x_1 divisible by $\Delta = 5$ without remainder, for

$u_1 = 50$ and $u_2 = 50$, for $\alpha < 50$,

$u_1 = \alpha + \Delta$ and $u_2 = \Delta \mathrm{int}(\frac{1}{2}(100 - \alpha)/\Delta)$, for $\alpha > 50$, where $\Delta = 5$ is the prominence level.

Obviously, the FDS and the PFS coincide for $\alpha < 50$.

THE PROMINENT DISTRIBUTION SCHEME

The *prominent distribution scheme* (PDS) only uses the insight that experimental subjects tend to choose prominent numbers in the area of the individually rational outcomes. It predicts all agreement outcomes on prominent numbers, which are individually rational.

Accordingly, the *prominent distribution scheme* (PDS) predicts the set of all grid points (x_1, x_2) with the properties $x_1 + x_2 = 100$, $x_i \geq u_i$ with $u_i = v(i)$, for $i = 1, 2$, and x_1 divisible by $\Delta = 5$ without remainder. Recall, that the value $v(i)$ denotes the alternative value of player i, $i = 1, 2$.

A good performance of this concept would mean that the attraction to prominent numbers is so strong that it eclipses other possible effects.

Notice, that only for the NAA the lower bound prediction depends on the alternative value of the opponent.

THE MEASURE OF PREDICTIVE SUCCESS

The comparison between the predictive power of the different distribution schemes for the experimental data under consideration will be based on the *measure of predictive success* (shortly: *success measure*) introduced by Selten and Krischker (1982). The success measure is the difference $S = H - A$ between the *hit rate* H and the *area* A. The hit rate H is the relative frequency of successful predictions. Two theories cannot be compared by their hit rates alone because one theory might predict a much larger region than the other one. Therefore, a measure for the relative size of the predicted region, the area A, is subtracted from the hit rate. The success measure of a theory is between -1 and 1. The higher the success measure is, the better is the predictive power of the theory.

A thorough justification of the success measure has be given in Selten (1991). There, it has been shown that among all measures based on hit rate and area the difference measure described above is singled out by desirable properties.

RESULTS OF THE COMPARISON OF THE PREDICTION CONCEPTS

For each gametype and for each of the six independent subject groups the success measure of each of the four prediction concepts was calculated. For the sake of simplicity we shall report theses data in the aggregated form of the average over the six independent subject groups. The gametypes with two high type players and $\alpha > 50$ are not evaluated, since (almost) no agreements were reached.

Table 6.11:   Average success measure

α	Gametype	NAA	FDS	PFS	PDS
30	(0, 0)	.774	.774	.774	.680
	(0,30)	$-.314$	.881	.881	.705
	(30, 0)	$-.280$	.819	.819	.657
	(30,30)	.854	.854	.854	.635
45	(0, 0)	.764	.764	.764	.639
	(0,45)	$-.390$	.716	.716	.607
	(45, 0)	$-.340$	.723	.723	.668
	(45,45)	.813	.813	.813	.540
55	(0, 0)	.282	.396	.668	.589
	(0,55)	.180	.318	.486	.377
	(55, 0)	.193	.350	.466	.391
60	(0, 0)	.209	.396	.616	.537
	(0,60)	.266	.451	.633	.534
	(60, 0)	.060	.510	.547	.425
70	(0, 0)	.388	.297	.680	.620
	(0,70)	.314	.449	.353	.245
	(70, 0)	.387	.506	.598	.510

NAA: *Negotiation Agreement Area*,   FDS: *Fair Distribution Scheme*,
PFS: *Prominent Fair Distribution Scheme*,   PDS: *Prominent Distribution Scheme*

For $\alpha < 50$ the FDS and the FPS coincide. For the cases where both players have the same alternative value these concepts also coincide with the NAA.

In order to detect statistically significant differences between the predictive power of the theories, a one-sided Wilcoxon matched-pairs signed-ranks test with the success measures for the six independent subject groups is conducted for each pairwise comparison of distribution schemes. The null hypothesis that there is no difference in the success measures of two distribution schemes is tested against the alternative that one of the theories has higher success measures. The results of these tests is summarized in table 6.12.

Table 6.12:   Ranks of performance and significance of performing best.

α	Gametype	NAA		FDS		PFS		PDS
30	(0, 0)	2	o	2	o	2	o	4
	(0,30)	4		1.5	•	1.5	•	3
	(30, 0)	4		1.5	•	1.5	•	3
	(30,30)	2	•	2	•	2	•	4
45	(0, 0)	2	•	2	•	2	•	4
	(0,45)	4		1.5	•	1.5	•	3
	(45, 0)	4		1.5	o	1.5	o	3
	(45,45)	2	•	2	•	2	•	4
55	(0, 0)	4		3		1	•	2
	(0,55)	4		3		1	•	2
	(55, 0)	4		3		1	o	2
60	(0, 0)	4		3		1	•	2
	(0,60)	4		3		1	o	2
	(60, 0)	4		2		1	•	3
70	(0, 0)	4		3		1	•	2
	(0,70)	3		1	o	2		4
	(70, 0)	4		3		1	o	2

NAA: *Negotiation Agreement Area,*   FDS: *Fair Distribution Scheme,*
PFS: *Prominent Fair Distribution Scheme,*   PDS: *Prominent Distribution Scheme*

Table 6.12 displays the rank of the distribution concept according to the average success measure. A " o " denotes the distribution scheme with the highest average success measure, which is not significantly better than the second best distribution scheme. A "•" denotes the distribution scheme with the best performance, which is significantly better than the second best distribution scheme (by the Wilcoxon matched-pairs signed-ranks test at a significance level of .025, one-sided).

Besides the cases were the NAA predicts the equal split it performs badly. The low (some-times even negative) success measure is caused by the fact that the lower bound for the strong player is too high. Contrary to games with complete information the strong player has no possibility to "prove" his strength and may always be seen as a weak player who pretends to be strong. This makes the strong player much weaker than in games with complete information and leads to the failure of the NAA for games with incomplete information.

The best performing prediction concept is the set of the prominent numbers in the fair distribution area. The concept performs significantly better than the FDS. This means that the players tend to agree at the prominent numbers in the fair distribution area. According to the investigations about the agreement outcomes (see Section 6.2) and the fact that over 70% of all numbers are prominent (see Section 6.3) this could be expected. Only once, the fair distribution scheme performs better, but not significantly.

The prominent distribution scheme is the second best in performance for $\alpha > 50$. This underlines the extraordinary importance of the prominent numbers. However, the prominence in combination with the fairness concept leads to the best prediction of the agreement outcomes.

It is the author's opinion that fair solutions are not only used because the subjects are concerned about fairness and justice, but because these values are focal points in the bargain-ing situation. If the fairness norms spread a range for bargaining, the subjects try to exploit this in their own favor. This could be seen in the asymmetric solutions for $\alpha > 50$. In the case of $\alpha < 50$ this bargaining range seems to be too small to be worthwhile to exploit. In view of the discount factor the subject seems to prefer a "quick" agreement on the focal point of the equal split.

6.7   AVERAGE PAYOFF AND EXPECTED PAYOFF IN EQUILIBRIUM

The Nash equilibria in pure strategies for the two person game with incomplete information were determined in Chapter 3. In a game playing experiment we cannot observe the
underlying strategies of the subjects. But, we can calculate the average payoff of a subject
in the four possible situations (strong first mover, weak first mover, strong second mover,
and weak second mover) and analyze whether these average payoffs are in the ranges of the
expected equilibrium payoffs. These ranges were determined in Section 3.3. Theorem 3
specifies these ranges for the case of discounting $(0 < \delta < 1)$. Table 6.13 gives the ranges of
the expected equilibrium payoffs for the different parameter values of α and $\delta = .99$.

Table 6.13:   Ranges for the expected payoffs in equilibrium for $\delta = .99$

Ranges for the expected payoffs in equilibrium				
α	$P_H^1 \in$	$P_H^2 \in$	$P_L^1 \in$	$P_L^2 \in$
30	[30,84.5]	[29.7,84.5]	[0,84.5]	[0,84.5]
45	[45,77]	[44.55,77]	[0,77]	[0,77]
55	[55,77]	[54.45,77]	[0,72]	[0,72]
60	[60,79.5]	[59.4,79.5]	[0,69.5]	[0,69.5]
70	[70,84.5]	[69.3,84.5]	[0,64.5]	[0,64.5]

The lower bounds of these ranges are induced by the individual rationality constraints.
Before we look at the distribution of the average payoffs of the subjects, we first investigate
whether the fundamental equilibrium condition of individual rationality holds in the game
playing experiment. Since it is trivially fulfilled for the low types we analyze whether the
high types were able to achieve average payoffs which at least respect the individual rationality. For each parameter value of α and each of the six independent subject groups the
average payoffs of the two high type players will be calculated. Table 6.14 gives the average
payoff of a high type first mover (P_H^1) and a high type second mover (P_H^2) in each of the six
independent subject groups. A "•" indicates that this value is not individually rational. For
these cases it can be excluded that the players play Nash equilibria in pure strategies.

Table 6.14: Average payoffs of the high types of the first and the second mover

α	Subject Group	P_H^1	P_H^2
30	1	38.55	39.04
	2	44.54	41.92
	3	39.25	38.98
	4	46.44	46.19
	5	46.72	45.79
	6	42.04	41.56
45	1	45.34	44.68
	2	44.71 •	45.86
	3	42.43 •	43.02 •
	4	44.18 •	44.63
	5	47.90	47.12
	6	45.26	44.88
55	1	54.84 •	54.66
	2	54.71 •	55.83
	3	54.47 •	56.28
	4	53.34 •	52.81 •
	5	53.63 •	54.38 •
	6	54.19 •	53.62 •
60	1	57.22 •	57.60 •
	2	60.55	60.33
	3	56.71 •	57.65 •
	4	56.52 •	52.74 •
	5	59.79 •	61.64
	6	57.13 •	56.30 •
70	1	68.95 •	69.70
	2	68.43 •	67.97 •
	3	68.27 •	67.16 •
	4	67.66 •	67.54 •
	5	66.00 •	65.52 •
	6	66.39 •	67.68 •

The result is striking. For the parameter values of $\alpha > 50$ only one subject group exceeded α as the average payoff for the high type of player 1. For the average payoff of the high type of the second mover the result is less extreme, but with the increase of α the violations of the pure strategy equilibrium prediction become clearer. The less extreme result for the high type of the second mover is due to the fact that the predicted lower bound is lower than for P_H^1 and that there is no significant difference between the expected payoff of the first and the second mover (see Section 6.2).

For $\alpha = 45$ only three sessions violate individual rationality for the high type of the first mover and we observe only one violation for the expected payoff of the second mover. For $\alpha = 30$ no violations are observable.

If the subjects use mixed strategies the average payoff is a realization of the expected payoff with respect to the probability distributions which control the mixing. A non-individually rational average payoff is not necessarily a contradiction to the assumption that Nash equilibria are played since the realization could be within the variance. If, however, the subjects use pure strategies, the expected payoff is no longer stochastic and therefore the average payoff coincides with the expected payoff. Each non-individually rational average payoff is a violation of the assumption that Nash equilibria in pure strategies are played. Since we cannot exclude that the subjects use mixed strategies we apply a statistical test in order to answer the question whether the observed average payoffs are "serious" violations. We shall use the Binomial test for the six independent subject groups.

For $\alpha = 55$, 60, and 70, respectively, the Binomial test rejects the null hypothesis that the average payoff of H_1 is equally likely to be below and above α at a significance level of .109 (one-sided). The alternative that the average payoff is below α is favored. For the high type second mover, only for $\alpha = 70$ the alternative of an average payoff below $\delta\alpha$ was favored by the Binomial test at a level of .109 (one-sided). For $\alpha = 30$ the one-sided Binomial test rejects the null hypothesis at a level of .016 (one-sided) in favor of the alternative that the players have average payoffs above the individual rationality payoff bound.

If the Binomial test rejects the hypothesis of average payoffs equally likely below and above the individual rationality bound in favor of the alternative of average payoffs below this bound we find a contradiction of the assumption that Nash equilibria are played.

It should be emphasized here, that a misconception of the subjects about their final payoff after the termination of a play was not possible. The discounted value was explicitly shown on the computer screen after the termination of the play (see Appendix A, figure A3). Therefore, the effect of non-individually rational payoffs cannot be explained by subjects thinking that they received the (non-discounted) outcome as payoff.

It is surprising that the individual rationality condition is violated so frequently, for $\alpha > 50$. In particular, this means that the strong first mover is, on average, not able to achieve the value α which he could guarantee himself by a unilateral break off. One might conjecture that the players learn from this experience and are able to achieve higher expected payoffs in the high experience levels. But, this is not the case. The average payoffs of the high types in the first experience level, as well as in the fourth experience level look very similar to the averages over all four experience levels. For the sake of brevity, they will not be reported here, but it should be mentioned that in the fourth experience level for $\alpha = 55$ there are 5 groups and for $\alpha = 60$ and 70 there are both 6 subject groups with average payoffs of H_1 violating individual rationality. For the high type second mover these are 5, for $\alpha = 55$ and 60, and 4 subject groups for $\alpha = 70$, respectively.

The average payoff lower than α and $\delta\alpha$ is explainable by the fact that the break off frequency is high for the plays of games with $\alpha > 50$ and even rises with the increase of α. But, these conflicts are mostly reached after a considerable number of steps. On the other hand, the surplus the strong player gains in case of agreement is not so large to compensate this loss. However, it will be shown in Section 7.4, that the players do not learn to break off immediately in case they are strong.

The following five figures (6.8 to 6.12) show the distribution of the average payoffs of the four possible types (H_1, H_2, L_1, and L_2) for each of the five parameter values of α, and moreover, the ranges of the expected equilibrium payoffs. For each parameter value of α we observed six independent subject groups with six subjects each. This means that we are able to calculate 36 average payoffs for each of the four types. They are represented by the stars in the figures. The horizontal lines represent the ranges of the expected equilibrium payoffs.

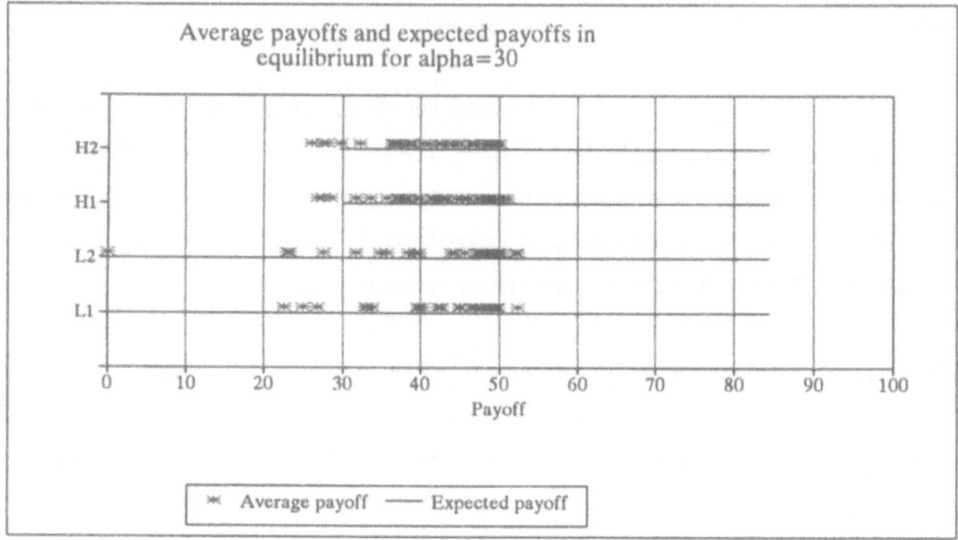

Figure 6.8:   Distribution of the average payoffs and the range of the expected equilibrium payoffs for $\alpha = 30$

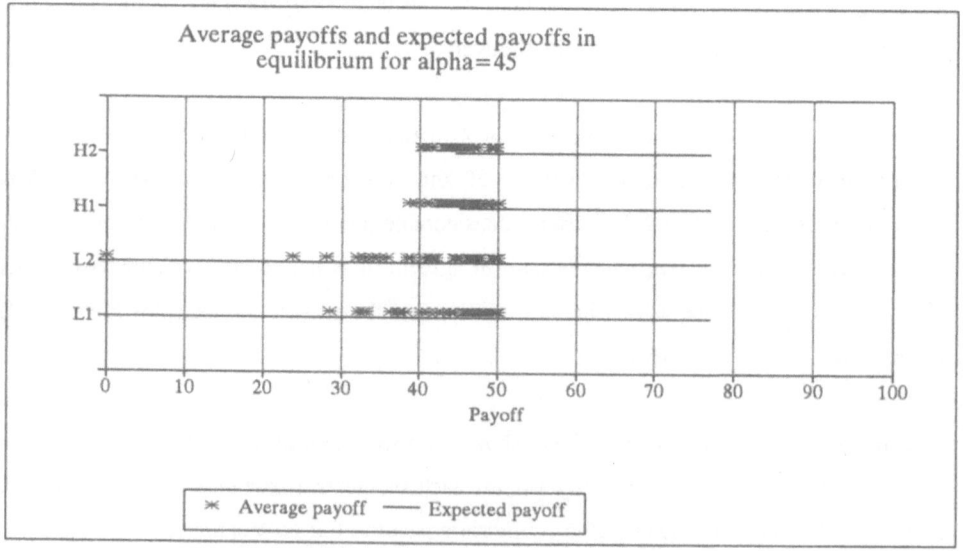

Figure 6.9:   Distribution of the average payoffs and the range of the expected equilibrium payoffs for $\alpha = 45$

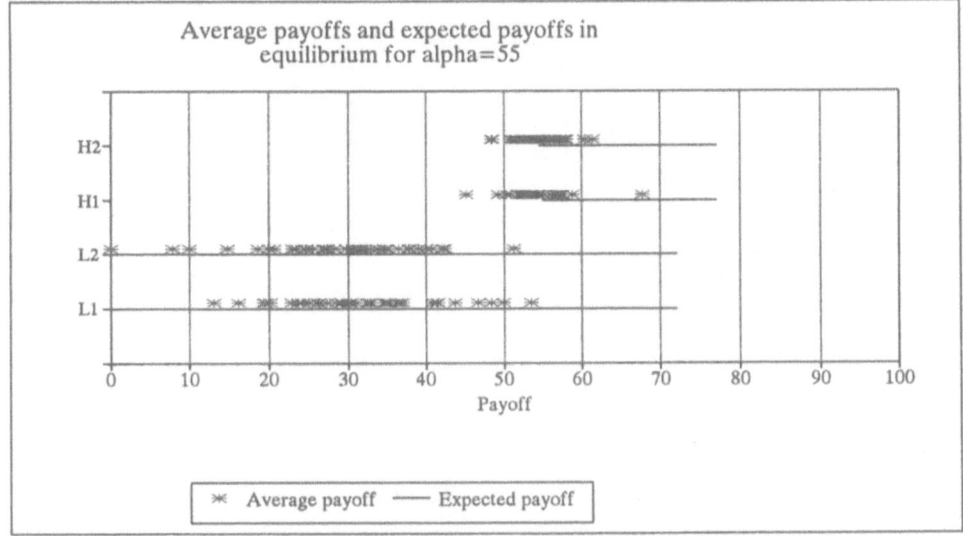

Figure 6.10:   Distribution of the average payoffs and the range of the expected equilibrium payoffs for $\alpha=55$

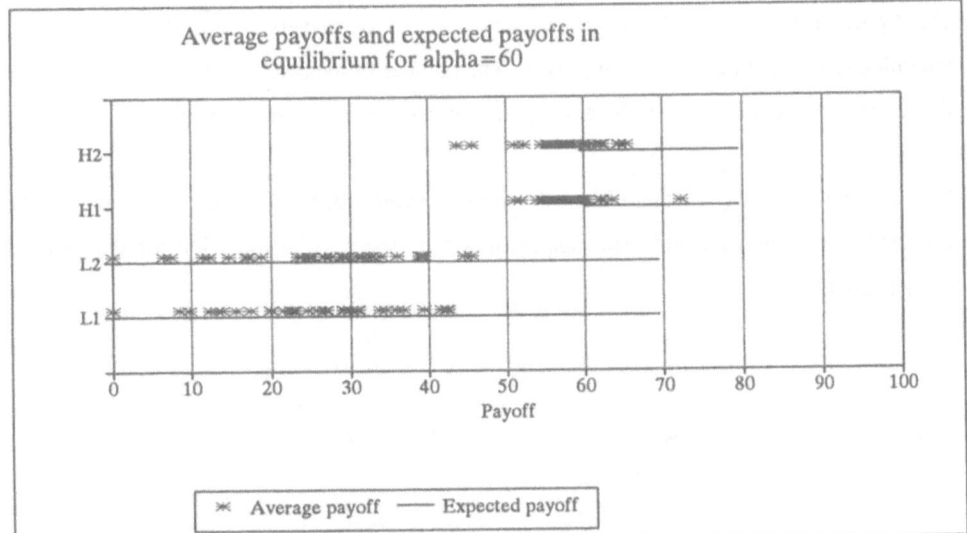

Figure 6.11:   Distribution of the average payoffs and the range of the expected equilibrium payoffs for $\alpha=60$

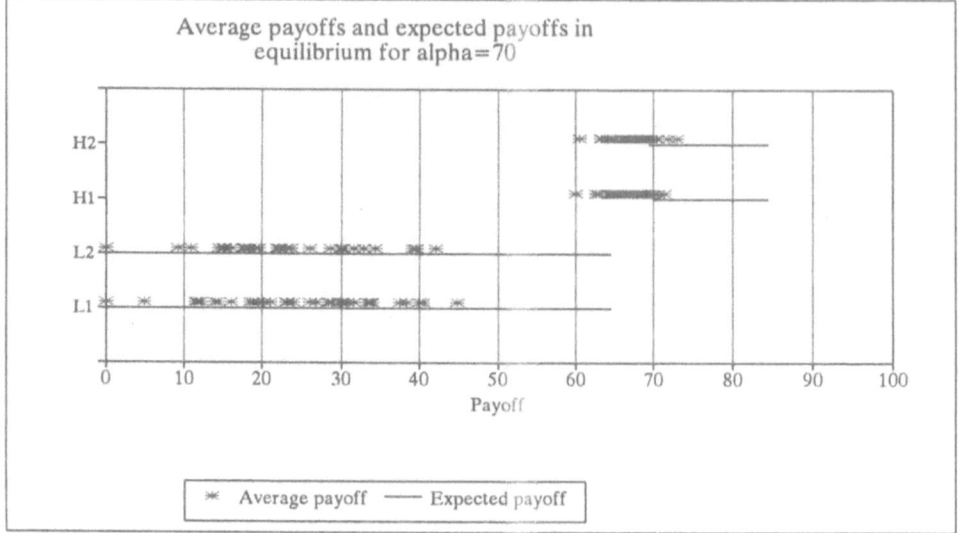

Figure 6.12:   Distribution of the average payoffs and the range of the expected equilibrium payoffs for $\alpha = 70$

The figures reflect again that a considerable number of players is not able to reach an individually rational payoff as a high type. The number of average payoffs located left from the range of the expected equilibrium payoffs increases with an increase in α.

For the low type players there are no average payoffs located outside the range of the expected equilibrium payoffs. The ranges, however, seem to be too wide for the realized average payoffs.

Chapter 7. Break Offs

Why do players break off bargaining games with a positive surplus? The analysis in this chapter examines this question by studying the special circumstances under which break off occurred and examines whether under the same conditions also agreements were possible.

7.1   Conflict Frequencies

At first we shall give an overview over the number of break offs in the different gametypes and experience levels. This more detailed description allows to study the change in the conflict frequencies over the different experience levels. Moreover, the aggregated conflict frequencies will be calculated for each parameter value of α.

Table 7.1 shows that break offs are a serious problem (in terms of efficiency losses) for the underlying bargaining game. Especially in plays of games with $\alpha < 50$ these numbers are seriously high since in all cases a profitable agreement was possible for both players and this was common knowledge.

For each parameter value of α, a Friedman two-way analysis of variance and an order test is conducted. Both two-sided tests do not reject the null hypothesis of no difference in the conflict rates in the four experience levels at a significance level of .2 for $\alpha=30$, $\alpha=55$, $\alpha=60$, and $\alpha=70$. Only for $\alpha=45$ the one-sided order test rejects the hypothesis at a level of .01 in favor of the alternative of a decreasing trend.

Table 7.1: Number of break offs in the plays of the different gametypes and experience
levels with 18 plays each

α	Gametype	Experience level			
		1	2	3	4
30	(0, 0)	0	0	1	0
	(0,30)	3	2	3	2
	(30, 0)	4	1	1	2
	(30,30)	7	6	7	2
45	(0, 0)	0	0	0	0
	(0,45)	8	5	1	1
	(45,0)	4	7	3	2
	(45,45)	13	14	10	10
55	(0, 0)	0	1	2	0
	(0, 55)	9	5	9	12
	(55, 0)	13	10	10	10
	(55,55)	18	18	18	18
60	(0, 0)	1	2	1	0
	(0,60)	13	12	10	15
	(60, 0)	11	14	13	10
	(60,60)	18	17	18	17
70	(0, 0)	0	2	1	0
	(0,70)	14	14	16	15
	(70, 0)	16	13	15	15
	(70,70)	18	18	18	18

Tables 7.2 through 7.6 provide the conflict frequencies for each value of α. The matrix
representation shows the four conflict frequencies between a first mover of type L or H and
a second mover of type L or H.

Interestingly, the conflict frequency between two low type players seems to be independent of α, and strictly below 1%. The phenomenon of break offs by weak players will be studied in Section 7.2.

With the exception of $\alpha=55$, it does not make any difference in the conflict frequencies whether the high type or the low type is the first mover.

Table 7.2: Conflict frequencies for $\alpha=30$

Second Mover

		L	H
	L	.01	.14
First Mover			
	H	.11	.31

Table 7.3: Conflict frequencies for $\alpha=45$

Second Mover

		L	H
	L	0	.21
First Mover			
	H	.22	.65

<u>Table 7.4:</u> Conflict frequencies for $\alpha = 55$

Second Mover

		L	H
	L	.04	.49
First Mover	H	.60	1

<u>Table 7.5:</u> Conflict frequencies for $\alpha = 60$

Second Mover

		L	H
	L	.06	.70
First Mover	H	.67	.97

<u>Table 7.6:</u> Conflict frequencies for $\alpha = 70$

Second Mover

		L	H
	L	.04	.82
First Mover	H	.82	1

7.2   Break Offs by the Weak Player

The next question we shall address concerns the player who initiates the conflict. Table 7.7 shows that in plays of a strong and a weak player the striking majority of break offs is caused by the strong player. The fact that it occurs that a weak player decides to break off, although it is always more profitable to reach an agreement, is remarkable and needs to be studied.

Table 7.7: Number of break offs by the weak player in the plays of games of a weak and a strong player with 18 plays in each cell

α	Gametype	Experience level			
		1	2	3	4
30	(0,30)	0	0	2	2
	(30, 0)	0	0	1	1
45	(0,45)	1	1	0	0
	(45,0)	0	1	0	0
55	(0,55)	0	1	0	3
	(55, 0)	0	0	0	0
60	(0,60)	1	0	1	1
	(60, 0)	0	1	0	1
70	(0,70)	1	1	0	0
	(70, 0)	0	0	0	1

Overall, it occurred in 32 out of 1080 plays that the weak player broke off the bargaining. The circumstances under which the break off occurred will be studied in the following.

Let b be the *maximal offer for the weak player who initiated the break off*. Let r be the *maximal offer to the other player*. By *maximal offer* we mean the maximal offer to this player during the whole bargaining game.

Table 7.8: Circumstances of break offs by the weak player

Circumstances of break off	# plays
b < 50	26
b = 50	5
b > 50	1
r < 50	15
r = 50	13
r > 50	4
b < 50 and r < 50	10
b ≥ 50 and r < 50	5
b < 50 and r ≥ 50	16
b ≥ 50 and r ≥ 50	1

The result is striking. In 26 out of 32 cases, the player who initiated the break off received a maximal offer of less than 50 and in 16 of these cases he himself made an offer of 50 or more to the opponent.

Apparently, the low type player wanted to obtain at least 50 and rather broke off than receiving less than 50. This phenomenon supports Bolton's fairness hypothesis (Bolton, 1991), which states that players are not only interested in the absolute payoff, but also in the relative payoff compared to the other player. If the relative difference between the offers is seen as too large, the offer is rejected as too unfair. The data show that players are willing to punish unfair behavior, even if they themselves suffer from this punishment.

7.3   BREAK OFFS BY THE STRONG PLAYER

THE GAMES WITH $\alpha > 50$

A player of type H faces with probability ½ an opponent of type H and he knows that they cannot agree on individually rational agreements for both players. Since the player does not know the type of his opponent, he must try to identify a strong opponent in order to break off the non-profitable bargaining. Under the assumption that a strong opponent makes no errors (and is not altruistic), he will never demand a value which is less than α and thereby will never offer an individually rational amount to the other player. However, this behavior might be imitated by the weak player in order to receive a high payoff in a play with another weak player. The question is whether a strong player knows to distinguish the imitation behavior from the behavior of a strong player and whether non-individually rational amounts offered are the only cause for the break offs.

The main point of interest seems to be the study of the behavior of the weak player that leads to a conflict. What happened before the break off of the strong player in a play of a strong and a weak player? In the following analysis we shall categorize the bargaining process into three different shapes. This categorization refers to the behavior of the weak player before the break off. We shall shortly speak of three different *shapes of behavior of the weak player*.

Shape A1: *The weak opponent did not reveal his type and did not offer an individually rational amount to the strong player.*

Shape A2: *The weak opponent revealed his type and did not offer an individually rational amount to the strong player.*

Shape A3: *The weak opponent revealed his type and offered an individually rational amount to the strong player.*

Remember, that *revealing* of the weak player means that he demands less than α for himself. Each amount of more than α is an individually rational amount for the strong player. A fourth possible shape which might be described by non-revealing but offering an individually rational amount to the strong player is not possible, since α is greater than 50. Each weak player's offer which yields an individually rational amount for the strong player reveals the type of the weak player.

An (individually rational) strong opponent can only behave according to shape A1. Shapes A2 and A3 reveal that the opponent is weak.

Figure 7.1 gives the number of occurrence for each of the three shapes in the different game-types.

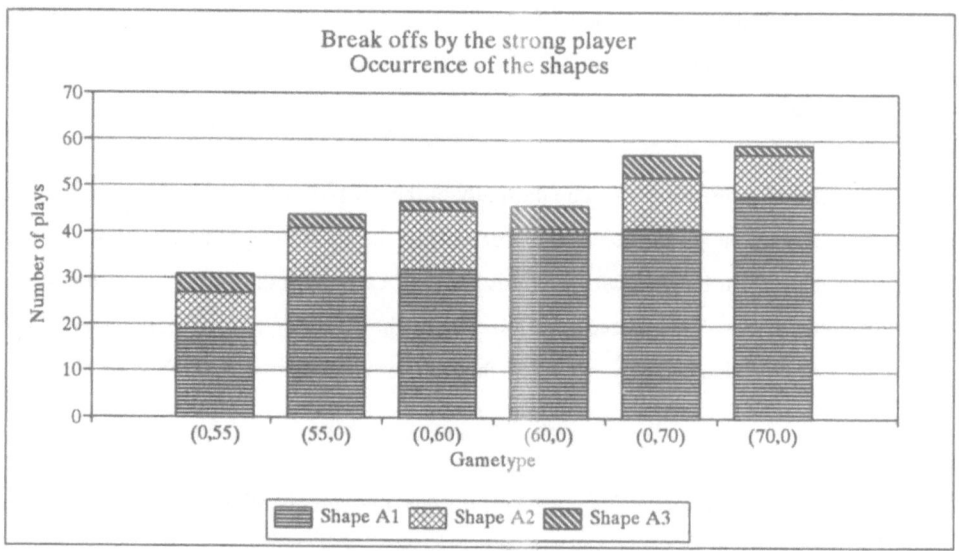

Figure 7.1:   Occurrence of shapes A1, A2, and A3

The first shape has the highest occupation, which means that in the majority of the cases the strong player has initiated a conflict in a play with a weak player who did not reveal his type and who did not offer an individually rational amount to the strong player.

For all three shapes we shall now study the circumstances of the break off in a more detailed way. We shall calculate the step of the bargaining in which the break off occurred and the size of the "gap" between the bargaining offers. There might be several ways to measure the *gap of disagreement* between the players. One possibility is to take the difference between each players last demand before the break off. However, this might be distorted by high demands serving as threats before the break off. Therefore, we shall measure the gap as the absolute value of the difference between the highest offer for the strong player made by the weak player and the lowest demand made by the strong player for himself during the bargaining process. In fact, these are the minimal demands each player has made for himself. For example, a minimal demand of 40 for the weak player and a minimal demand of 80 for the strong player would lead to a gap of 20 (resulting from $|60-80|$). The idea is that these values are the guidelines for the bargaining, although the players are not committed to previous demands.

The following three tables (7.9 to 7.11) present the average gap and the average step of break off for the three shapes. For the sake of lucidity we shall aggregate the low and the high experience levels.

Table 7.9: Average gap and average number of steps for shape A1

α	Gametype	Experience level 1 & 2			Experience level 3 & 4		
		# plays	Avg gap	Avg steps	# plays	Avg gap	Avg steps
55	(0,55)	9	44.78	3.44	10	37.40	4.60
	(55, 0)	14	32.57	4.14	15	34.40	7.00
60	(0,60)	16	49.06	6.13	16	55.31	6.63
	(60, 0)	20	53.55	6.60	20	51.90	8.65
70	(0,70)	21	60.95	4.90	20	63.80	4.30
	(70, 0)	24	59.71	6.33	24	57.96	4.88

Table 7.9 shows a tremendous difference between the minimal demands of the players. The greater α is the larger is the gap. From the high demands of the opponent the strong player seems to conclude that he faces a strong opponent and breaks off the bargaining after roughly 5 steps. This result fits nicely into the insights to be gained from the examination of the revelation behavior (see Section 8.4). There we shall find that the median step of revelation is step 4. If all players would behave according to this median, a strong player knows in step

5 that he is playing a strong opponent and can do no better than breaking off.

Table 7.10:   Average gap and average number of steps for shape A2

α	Gametype	Experience level 1 & 2			Experience level 3 & 4		
		# plays	Avg gap	Avg steps	# plays	Avg gap	Avg steps
55	(0,55)	4	26.75	7.50	4	16.00	9.50
	(55, 0)	6	14.50	5.67	5	14.20	12.40
60	(0,60)	7	22.43	9.86	6	20.00	27.67
	(60, 0)	1	25.00	6.00	0	–	–
70	(0,70)	5	31.60	11.00	6	26.33	12.33
	(70, 0)	4	33.25	4.00	5	28.40	14.00

In the behavior classified as shape A2 the weak player reveals his type by demanding less than α but makes on the other hand no offer that the strong player can accept (this means which yields him more than his alternative value). The gap between the demands is significantly smaller than in the shape A1 behavior but it also increases with α. The bargaining proceeds on average longer than in the previous shape. In the investigations on the revelation behavior we shall see that the median step of revelation is step 4. Very roughly speaking, in the low experience levels the strong player waits about 4 steps in addition to see whether the revealer makes a (for the strong player) acceptable demand and breaks off afterwards. In the high experience levels the bargaining proceeds significantly longer, for 10 to 28 steps.

The behavior of shape A2 only makes sense if the weak player is sure that the opponent is weak, too. If the opponent is strong he knows that they cannot agree on his demands and the revelation was of no use. A closer look at the actual offers shows that in the majority of the plays (39 out of 53) the weak player revealed his type by proposing 50 and this was also the maximal offer to the opponent. In 8 cases the weak player's maximal offer was less than 50, in 3 cases it was between 50 and α, and in 3 cases it was exactly α.

Apparently, the majority of the break offs of this shape is explained by a weak player who reveals his type by proposing 50:50 and sticks to this demand (strongly believing that the other player is weak, too).

Table 7.11:   Average gap and average number of steps for shape A3

α	Gametype	Experience level 1 & 2			Experience level 3 & 4		
		# plays	Avg gap	Avg steps	# plays	Avg gap	Avg steps
55	(0,55)	0	–	–	4	8.50	16.50
	(55, 0)	3	12.67	22.67	0	–	–
60	(0,60)	1	19.00	15.00	1	14.00	19.00
	(60, 0)	3	5.33	26.00	2	11.50	15.00
70	(0,70)	0	–	–	5	6.20	11.00
	(70, 0)	1	18.00	14.00	0	–	–

Table 7.11 reflects the case that the weak player revealed his type and made acceptable offers to the strong player. Nevertheless, the strong player broke off the negotiations. The number of occurrences is small, but it happens with remaining gaps of approximately 10. The number of steps is overproportionally long in these cases.

In 8 of the 20 cases the maximal offer was less than $\alpha+5$ and in 4 cases it was exactly $\alpha+5$. As already noticed in the study of the agreement outcomes (see Section 6.4), a player seems to strive for at least one prominence level (and not one money unit) in addition to his alternative value. In 8 cases the maximal offer of the weak player exceeded α by more than 5. Here, the average gap was 5. This phenomenon might be explained by stubborn bargaining partners. A look at the individual player level showed that these results are caused by different players.

The observations concerning the behavior classified as shape A1 suggested that the strong player breaks off because the low type's behavior is not distinguishable from a high type's behavior. However, this means that the statistic over the plays of two strong players should look very similar to the shape A1 behavior. Table 7.12 shows that this is indeed the case.

Table 7.12:   Average gap and average number of steps in the bargaining of two strong
               players that ended with conflict

α	Gametype	Experience level 1 & 2			Experience level 3 & 4		
		# plays	Avg gap	Avg steps	# plays	Avg gap	Avg steps
55	(55,55)	36	42.22	3.94	36	38.08	5.78
60	(60,60)	35	56.63	5.20	35	59.49	5.49
70	(70,70)	36	64.53	4.19	36	64.50	4.31

The size of the average gap as well as the average number of steps have the same magnitude as in bargaining games with a weak opponent who did not reveal his type and did not make an acceptable offer for the strong player.

An evaluation of the size of the gap at step 5 (the median break off step in A1) for shape A2 and shape A3 shows that it is significantly smaller than the gap in shape A1 and the plays of two strong players.

To summarize the break off behavior of the strong player for $\alpha > 50$, we find a critical phase of the game which is close to step 4 or 5. There, the strong player makes up his mind about the beliefs of the opponent's type. In the case that the gap between the demands is large (which means about 40, 50, and 60 for $\alpha = 55$, 60, and 70, respectively), the strong player decides to break off the negotiations. However, in case of a revelation of the opponent, which occurs approximately in step 4, the bargaining proceeds. If the revealer continues in making no acceptable offer (mostly the equal split), the strong player will break off. In cases where the weak player makes acceptable offers, the majority of maximal offers the strong player received was less or equal to $\alpha + 5$. In rare cases, where the strong player receives an offer greater than $\alpha + 5$ and nevertheless breaks off the remaining gap is small (on average 5) and the bargaining is overproportionally long.

THE GAMES WITH $\alpha < 50$

In the following we shall investigate the circumstances for break offs of the strong player for plays of games with $\alpha < 50$. Recall, that for all possible gametypes profitable agreements for both players are possible. Nevertheless, we observe a high proportion of conflicts which increases with the shrinkage of the surplus.

Like in the investigations on the break off circumstances for $\alpha > 50$ we shall distinguish shapes of the bargaining behavior of the weak player before the break off. The first distinction will concern the question whether or not the strong player received an individually rational offer. The second distinction will concern the minimal demand of the weak player. Speaking of a revelation of the weak player would mean observing a demand of less than α for the weak player. Actually, this never happened. However, we shall focus on the size of the highest offer the strong player received during the bargaining process. The offer of 50 will be chosen as a focal point. The following four shapes can be distinguished.

Shape B1: *The high type player received no individually rational offer*

Shape B2: *The high type player received an individually rational offer and the maximal offer was less than 50*

Shape B3: *The high type player received an individually rational offer and the maximal offer was equal to 50*

Shape B4: *The high type player received an individually rational offer and the maximal offer was greater than 50*

Figure 7.2 shows the occurrences of the shapes in the different gametypes. The following four tables (7.13 to 7.16) give the number of occurrence, the average gap (defined as before) and the average bargaining length in the low and in the high experience levels for each of the four shapes.

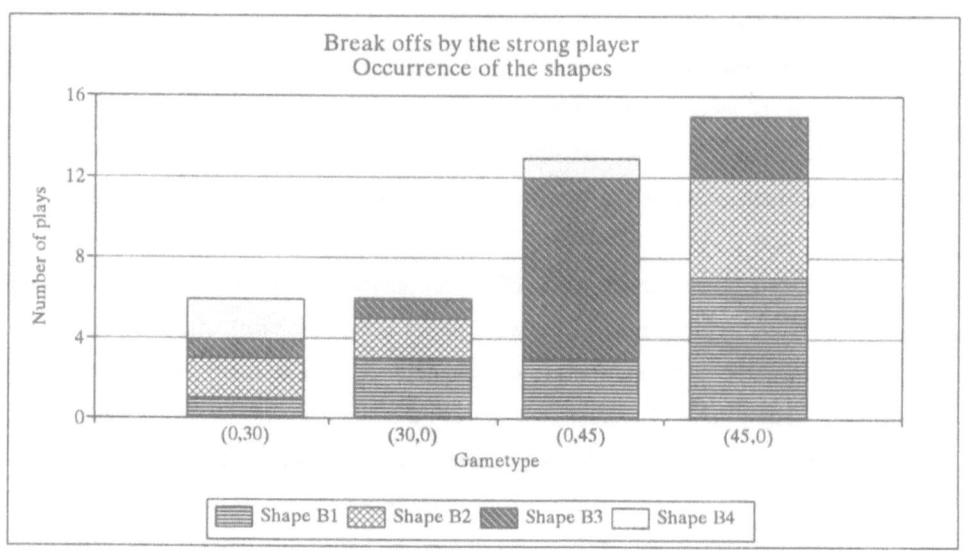

<u>Figure 7.2:</u>   Occurrence of shapes B1, B2, B3, and B4

<u>Table 7.13:</u>   Average gap and average number of steps for shape B1

α	Gametype	Experience level 1 & 2			Experience level 3 & 4		
		# plays	Avg gap	Avg steps	# plays	Avg gap	Avg steps
30	(0,30)	1	21.00	3.00	0	–	–
	(30, 0)	3	35.00	10.00	0	–	–
45	(0,45)	2	42.00	9.00	1	43.00	5.00
	(45, 0)	4	28.25	11.50	3	49.33	3.00

<u>Table 7.14:</u>   Average gap and average number of steps for shape B2

α	Gametype	Experience level 1 & 2			Experience level 3 & 4		
		# plays	Avg gap	Avg steps	# plays	Avg gap	Avg steps
30	(0,30)	2	14.00	13.00	0	–	–
	(30, 0)	2	16.50	12.00	0	–	–
45	(0,45)	0	–	–	0	–	–
	(45, 0)	4	6.50	8.00	1	14.00	8.00

Table 7.15:   Average gap and average number of steps for shape B3

α	Gametype	Experience level 1 & 2			Experience level 3 & 4		
		# plays	Avg gap	Avg steps	# plays	Avg gap	Avg steps
30	(0,30)	1	19.00	13.00	0	–	–
	(30, 0)	0	–	–	1	15.00	24.00
45	(0,45)	8	14.62	14.00	1	10.00	21.00
	(45, 0)	2	7.00	25.00	1	2.00	12.00

Table 7.16:   Average gap and average number of steps for shape B4

α	Gametype	Experience level 1 & 2			Experience level 3 & 4		
		# plays	Avg gap	Avg steps	# plays	Avg gap	Avg steps
30	(0,30)	1	20.00	13.00	1	2.00	9.00
	(30, 0)	0	–	–	0	–	–
45	(0,45)	1	5.00	29.00	0	–	–
	(45, 0)	0	–	–	0	–	–

For $\alpha=30$, as well as for $\alpha=45$ the fourth shape, in which the high type receives an offer greater than 50 seems to be of no relevance. In two of the cases the gap is very small (2 and 5). A look at the individual level shows that the two break offs for (0,30) are caused by different players with different opponents.

Overall, it is observable that except for the first shape only single break offs occur in the high experience levels.

For $\alpha=30$ the shapes B1 and B2 are most relevant. In the circumstances of shape B1 the high type did not receive an individually rational offer and broke off before step 10, on average. The average gap between 21 and 35 was quite large. The four break offs of shape B2 occurred after the strong player had received an offer between 31 and 49 (boundaries included). The average gap was approximately 15 and the average number of steps between

12 and 13.

For $\alpha=45$ figure 7.2 shows a difference in the distribution of the shapes for the two gametypes. If the weak player is the first mover (gametype (0,45)), the majority of break offs by the strong player occurred in the situation of shape B3 and if the strong player moves first (gametype (45,0)), shape B1 is observed most frequently. This means we observe a considerable difference, contingent on the first moving player. This is quite interesting since all previous investigations showed that the mover sequence does not have a significant influence on the bargaining. How can we explain this phenomenon? A possible explanation for the case that the weak player moves first is that he initially demands 50 and then remains at this demand. This is true for four of the nine cases subsumed under shape B3. This means that the weak player starts with a demand which leaves an individually rational amount to the strong player and during the bargaining process the weak player makes no "better" offer to the strong player. But, the strong breaks off since he strives for a higher agreement outcome.

If, however, the strong player moves first in the majority of break offs the strong player did not receive an individually rational offer (shape B1) or a maximal offer less than 50 (shape B2). An explanation is that the weak player "repeats" the high demands of the strong opponent (presumably thinking that the opponent is a "bluffing" weak player). The data show that in the cases subsumed under shape B1 and B2 the first demands of the second moving weak player are between 60 and 99, which is higher than the initial demand of the weak player as a first mover.

The observations can be roughly summarized by stating that for $\alpha=30$ the high type player breaks off because he did not receive either an individually rational offer or an offer of at least 50. For $\alpha=45$, however, the strong type mostly breaks off because he received no individually rational offer (in the case he is the first mover) or he receives no offer strictly greater than 50 (if he is the second mover). For both parameter values of α the number of conflicts decreases in the high experience level.

Since α is less than 50 a profitable agreement is also possible in plays of two high type players. However, they often fail to reach an agreement. Table 7.17 gives a global overview.

Table 7. 17:   Average gap and average number of steps in the bargaining of two strong
                players that ended with conflict

α	Gametype	Experience level 1 & 2			Experience level 3 & 4		
		# plays	Avg gap	Avg steps	# plays	Avg gap	Avg steps
30	(30,30)	13	13.69	16.08	9	19.89	17.33
45	(45,45)	27	21.30	8.85	20	13.25	10.20

The circumstances of the break offs will be studied in the following. We shall distinguish
four possible shapes of the bargaining process before the conflict occurred.

Shape C1: *None of the two players received an individually rational offer*

Shape C2: *Both players received an individually rational offer*

Shape C3: *The player who initiated the conflict did not receive an individually rational offer,
while the opponent did*

Shape C4: *The player who initiated the conflict received an individually rational offer, while
the opponent did not*

The occurrences of these shapes are displayed in the figure 7.3. The number of occurrences,
the average gap and the average length of bargaining in the low and in the high experience
phases is studied in the next four tables (7.18 to 7.21).

Table 7.18:   Average gap and average number of steps in the bargaining of two strong
               players that ended with conflict for shape C1

α	Gametype	Experience level 1 & 2			Experience level 3 & 4		
		# plays	Avg gap	Avg steps	# plays	Avg gap	Avg steps
30	(30,30)	0	–	–	1	90.00	2.00
45	(45,45)	6	33.00	8.00	2	37.50	4.50

Table 7.19:   Average gap and average number of steps in the bargaining of two strong
                players that ended with conflict for shape C2

α	Gametype	Experience level 1 & 2			Experience level 3 & 4		
		# plays	Avg gap	Avg steps	# plays	Avg gap	Avg steps
30	(30,30)	12	12.33	16.92	7	7.00	20.57
45	(45,45)	4	3.00	8.75	3	1.33	12.33

Table 7.20:   Average gap and average number of steps in the bargaining of two strong
                players that ended with conflict for shape C3

α	Gametype	Experience level 1 & 2			Experience level 3 & 4		
		# plays	Avg gap	Avg steps	# plays	Avg gap	Avg steps
30	(30,30)	1	30.00	6.00	1	40.00	10.00
45	(45,45)	16	20.37	9.38	15	12.40	10.53

Table 7.21:   Average gap and average number of steps in the bargaining of two strong
                players that ended with conflict for shape C4

α	Gametype	Experience level 1 & 2			Experience level 3 & 4		
		# plays	Avg gap	Avg steps	# plays	Avg gap	Avg steps
30	(30,30)	0	–	–	0	–	–
45	(45,45)	1	39.00	6.00	0	–	–

For $\alpha=30$ nearly all of the break offs occurred in the situation of shape C2. Both players
received individually rational offers, but they could not agree on the distribution of the
surplus. The bargaining was considerably long and the gap between the maximal offers
roughly 10, which equals one quarter of the surplus.

For $\alpha=45$ the majority of break offs was caused by a player who had not received an
individually rational offer up to step 10, but had made an individually rational offer to the

opponent. This happens in both, the low and the high experience level with roughly the same frequency and the same number of steps. In the high experience level, the gap is considerably smaller than in the low experience levels. Recall, that two high type players have only a surplus of 10.

Another source of break off is that both players received individually rational offers, but they cannot agree on the split of the surplus. The average gap is very small, which leads more to the conclusion of stubborn players than of one player claiming the whole surplus.

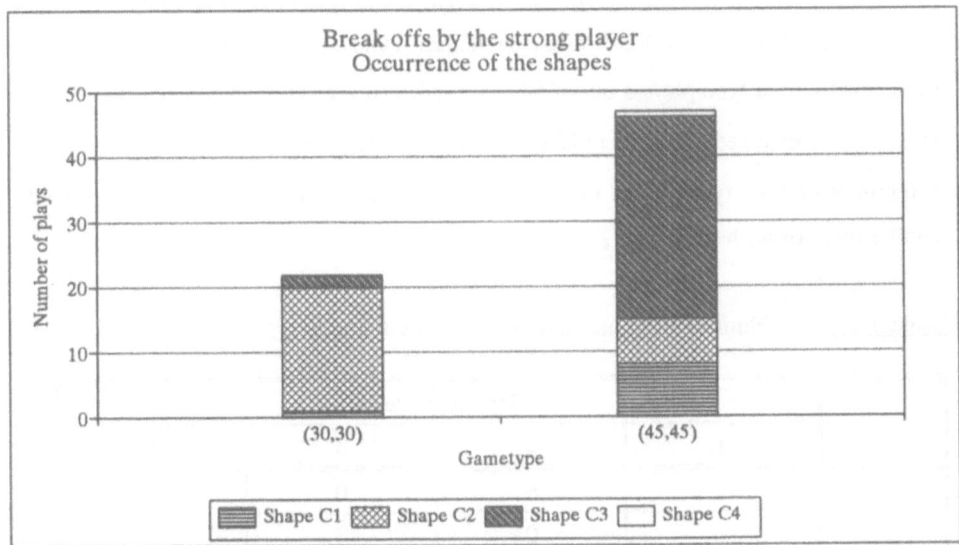

Figure 7.3:   Occurrence of shapes C1, C2, C3, and C4

7.4   IMMEDIATE BREAK OFFS

Especially for $\alpha > 50$ a high type player might think that the bargaining is not worth
while, since the surplus is too small and the alternative value is discounted with each
demand. As a consequence a high type player might break off at his first move (either as
first mover or as responder).

For $\alpha < 50$ only two immediate break offs happened. For $\alpha > 50$, the number of immediate
break offs is also very small. Due to the poorness of the data, we shall aggregate the data
over the six independent subject groups and the gametypes, but differentiate due to the
experience level. Hence, each cell of the subsequent table contains the number of immediate
break offs by high type players out of 54 plays ($54 = 6$ groups $\cdot$ 3 gametypes where a high
type player is involved $\cdot$ 3 parallel plays of each gametype) and with 72 high type players ($72
= 6$ groups $\cdot$ 4 high type players in the 3 gametypes where a high type player is involved $\cdot$ 3
parallel plays of each gametype).

Table 7.22:   Number of immediate break offs by the high type

α	Experience level			
	1	2	3	4
55	2	5	0	3
60	2	0	2	1
70	3	3	6	3

The table indicates no learning effect of the players concerning the immediate break offs.
The occupation of the cells in each experience level is too poor to confirm this observation
by a statistical test.

Hence, in case a high type is involved in a play, he breaks off immediately in 3% ($\alpha=55$),
in 2% ($\alpha=60$), or in 5% ($\alpha=70$) of all plays in his first decision step.

These numbers are very low, which means that the vast majority of strong players decided
to enter the bargaining.

7.5   Agreements under similar Circumstances

The Games with $\alpha > 50$

The study of the break off behavior of the strong player has shown that the main reason for break offs was the non-existence of an individually rational offer by the opponent. The strong player seemed to consider the bargaining as hopeless and broke off. Did it, however, happen that the weak player accepted the proposal of a strong player under the same circumstances? This means, we shall look at bargaining games where the weak player made no individually rational offer to the strong player and accepted all of a sudden. In analogy to the shapes A1 and A2 we shall define the following shapes.

Shape A1':   *The weak player did __not__ reveal his type and did __not__ offer an individually rational amount to the strong player, but accepted an offer of the strong player.*

Shape A2':   *The weak player revealed his type and did __not__ offer an individually rational amount to the strong player, but accepted an offer of the strong player.*

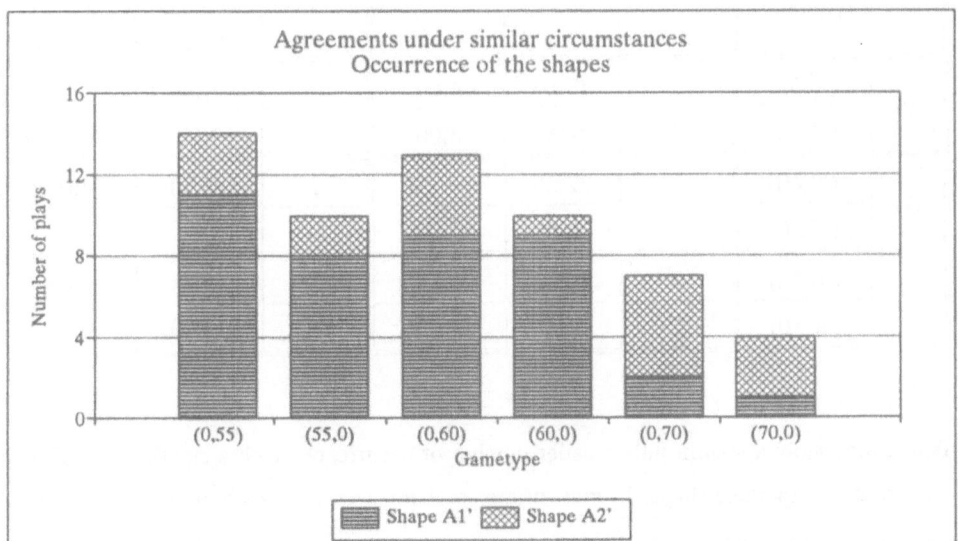

Figure 7.4:   Occurrence of shapes A1' and A2'

The following two tables (7.23 and 7.24) show the number of occurrences, the average gap, and the average length of bargaining for the low and the high experience levels for the shapes A1' and A2'.

Table 7.23:   Average gap and average number of steps for shape A1'

α	Gametype	Experience level 1 & 2			Experience level 3 & 4		
		# plays	Avg gap	Avg steps	# plays	Avg gap	Avg steps
55	(0,55)	8	27.37	2.50	3	24.00	7.33
	(55, 0)	3	28.33	4.33	5	22.80	7.40
60	(0,60)	5	44.40	9.60	4	47.75	4.50
	(60, 0)	6	46.50	3.67	3	38.33	6.33
70	(0,70)	1	50.00	10.00	1	48.00	8.00
	(70, 0)	0	–	–	1	75.00	5.00

Table 7.24:   Average gap and average number of steps for shape A2'

α	Gametype	Experience level 1 & 2			Experience level 3 & 4		
		# plays	Avg gap	Avg steps	# plays	Avg gap	Avg steps
55	(0,55)	1	15.00	4.00	2	24.50	3.00
	(55, 0)	1	8.00	3.00	1	7.00	15.00
60	(0,60)	2	22.50	11.00	2	22.50	6.00
	(60, 0)	1	25.00	5.00	0	–	–
70	(0,70)	2	28.00	11.00	3	36.33	2.67
	(70, 0)	2	5.00	9.00	1	30.00	9.00

Both tables show a significantly smaller number of occurrences with a smaller average gap than in the comparable shape. In most of the cases the average bargaining length is longer.

THE GAMES WITH $\alpha < 50$

For $\alpha < 50$ the two major reasons for the break off of the strong player were the non-existence of an individually rational offer and a maximal offer of not more than 50. Under the circumstance that no individually rational amount is offered to the strong player we shall study the plays in which the weak player accepted all of a sudden a proposal of the strong player. In the case that the weak player offers an individually rational amount to the strong player, but it was less or equal to 50, we shall investigate the plays in which the strong player accepted nevertheless. This leads to the definition of analogous shapes.

Shape B1':   *The weak player did <u>not</u> offer an individually rational amount to the strong player, but accepted an offer of the strong player.*

Shape B2':   *The weak player offered an individually rational amount to the strong player and the maximal offer was less than 50 and the strong player accepted this offer*

Shape B3':   *The weak player offered an individually rational amount to the strong player and the maximal offer was equal to 50 and the strong player accepted this offer*

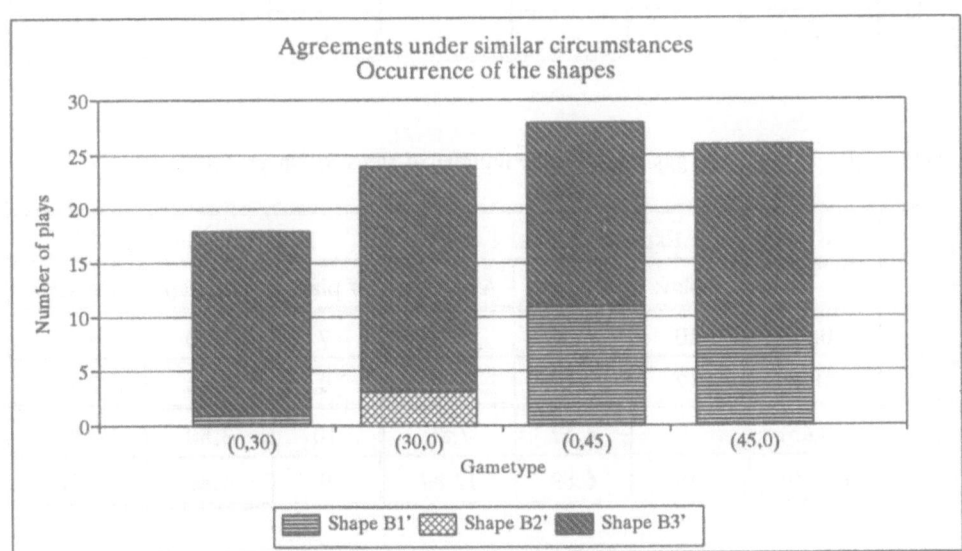

Figure 7.5:   Occurrence of shapes B1', B2', and B3'

The following three tables (7.25 to 7.27) show the statistics for these shapes.

Table 7.25:   Average gap and average number of steps for shape B1'

α	Gametype	Experience level 1 & 2			Experience level 3 & 4		
		# plays	Avg gap	Avg steps	# plays	Avg gap	Avg steps
30	(0,30)	1	19.00	36.00	0	–	–
	(30, 0)	0	–	–	0	–	–
45	(0,45)	5	16.40	4.80	6	10.00	2.67
	(45, 0)	7	14.71	8.43	1	5.00	3.00

Table 7.26:   Average gap and average number of steps for shape B2'

α	Gametype	Experience level 1 & 2			Experience level 3 & 4		
		# plays	Avg gap	Avg steps	# plays	Avg gap	Avg steps
30	(0,30)	0	–	–	0	–	–
	(30, 0)	2	1.50	15.00	1	1.00	10.00
45	(0,45)	0	–	–	0	–	–
	(45, 0)	0	–	–	0	–	–

Table 7.27:   Average gap and average number of steps for shape B3'

α	Gametype	Experience level 1 & 2			Experience level 3 & 4		
		# plays	Avg gap	Avg steps	# plays	Avg gap	Avg steps
30	(0,30)	10	6.30	7.40	7	3.00	6.43
	(30, 0)	12	9.92	6.50	9	4.44	7.33
45	(0,45)	7	4.57	7.86	10	3.80	6.80
	(45, 0)	9	6.89	12.89	9	4.89	4.22

For $\alpha=45$ shape B1' has a high occupation, but in all cases the average gap is significantly smaller than in the case of the strong player's break off.

For the plays subsumed under shape B3' the strong player accepted a maximal offer of 50, in contrast to shape B3, where the strong player broke off in the same situation. Table 7.15 already showed that in the high experience levels break offs in situations of shape B3 are rather seldom (only twice) in contrast to the low experience levels (ten times). Now, table 7.27 shows that, especially in the high experience levels, a maximal offer of 50 is more frequently accepted by the strong player (19 times in the high experience levels).

In the study of the bargaining process in Section 9.4 we shall come back to agreements of this kind, which will be called *sudden acceptance*. In this case the bargaining process suddenly ends by acceptance, although the two demand processes did not converge.

Chapter 8. Further Aspects

8.1   The Initial Demands

In the bargaining game each new demand that has to be made before an agreement is reached causes a discounting of the alternatives and the coalition value. To avoid non-necessary discounting steps a player should select each demand carefully. This is especially true for the initial demand. If it is accepted, no discounting occurs.

This section studies the distribution of the initial demands, and especially their development in the different levels of experience. For each of the five values of α, we shall present the distribution of the initial demand made by a low type player and those made by a high type player. For the sake of simplicity the experience levels 1 and 2, and the levels 3 and 4 are aggregated in the following figures. Furthermore the immediate acceptance rate of the equal split proposal will be investigated for the plays of games with $\alpha < 50$.

The games with $\alpha < 50$

For $\alpha = 30$ the initial demands in the low experience level are scattered in the range from 40 to 90 with major peaks on the prominent numbers. In the plays of the high experience level the striking majority of first demand is on the equal split. The one-sided order test rejects the null hypothesis that there is no trend in the initial demands in favor of the alternative of a decreasing trend at a significance level of .01.

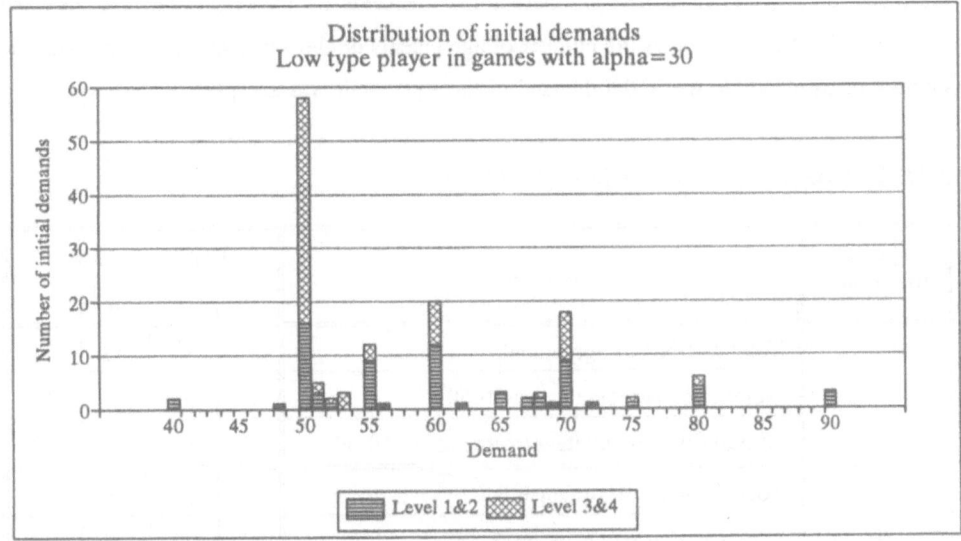

Figure 8.1: Distribution of the initial demands of a low type player at the different experience levels for $\alpha = 30$

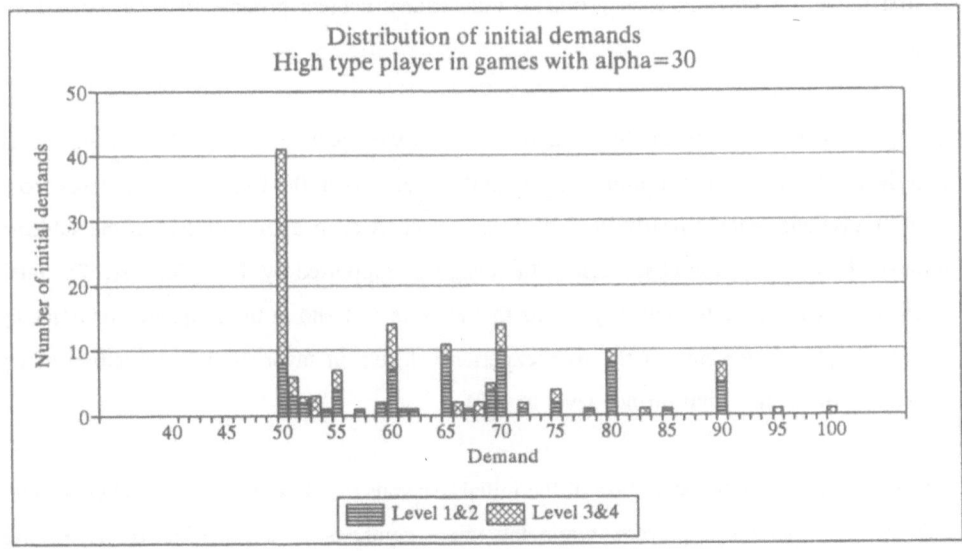

Figure 8.2: Distribution of the initial demands of a high type player at the different experience levels for $\alpha = 30$

However, it is not obvious that all these initial demands on the equal split are accepted immediately. Table 8.1 shows the number of agreements on the equal split and moreover the number of cases where the initial demand of the equal split was accepted.

Table 8.1: Agreements on the equal split for $\alpha=30$

Type of first mover	$\alpha = 30$	Experience level			
		1	2	3	4
L	Total number of agreements	33	34	32	34
	Number of agreements on 50:50	20	31	29	30
	Number of immediate agreements on 50:50	2	11	13	18
H	Total number of agreements	25	29	28	32
	Number of agreements on 50:50	20	22	25	28
	Number of immediate agreements on 50:50	0	3	9	16

The maximal number of possible agreements is 36 (=2 gametypes where this type of player is first mover · 3 plays in each group · 6 independent subject groups) in each experience level.

The table shows the large number of agreements on the equal split (over 87% in the fourth experience level). The proportion of equal split agreements in the first round increases from level to level and reaches nearly ½ in the fourth level. A clear learning effect of the subjects towards the equal split is observable. This insight is supported by the order test. The one-sided order test rejects the null hypothesis that there is no trend in the frequency of immediate equal split agreements in the four experience levels in favor of the alternative of an increasing trend, at a significance level of .001.

For $\alpha=45$ the major concentration of the initial demands of a low type player is between 50 and 60 for experienced players, while the initial demands of unexperienced players are scattered between 48 and 100. The one-sided order test rejects the null hypothesis that there is no trend in the initial demands in favor of the alternative of a decreasing trend at a significance level of .01.

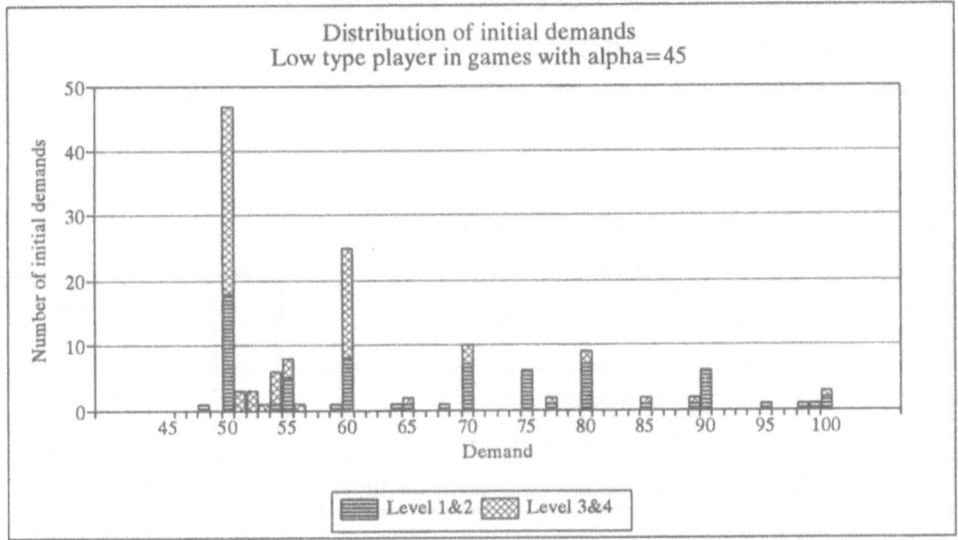

<u>Figure 8.3:</u>   Distribution of the initial demands of a low type player at the different experience levels for $\alpha = 45$

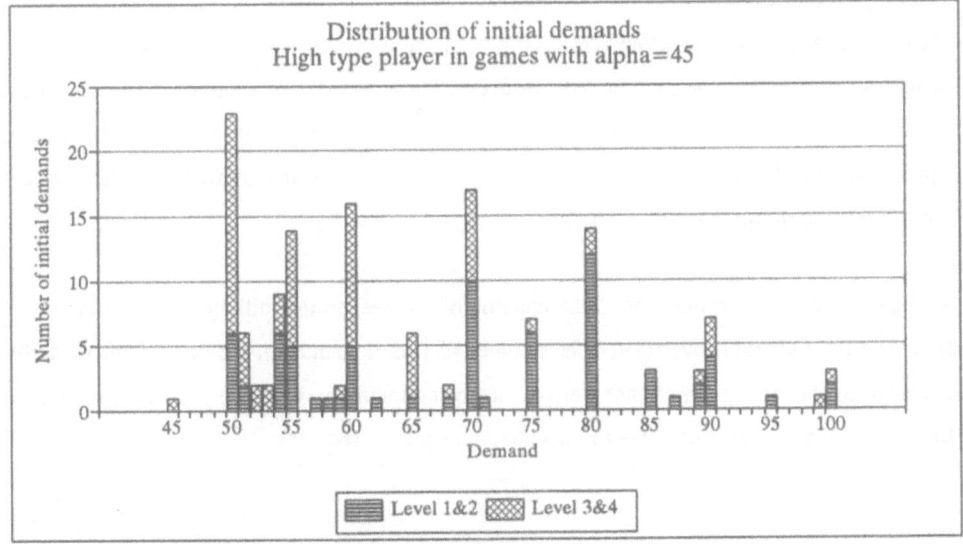

<u>Figure 8.4:</u>   Distribution of the initial demands of a high type player at the different experience levels for $\alpha = 45$

Table 8.2 provides an overview over the equal split agreements, especially those that are proposed in the first step and accepted immediately. As the figures already suggested, this proportion will not be as high as in the $\alpha=30$ case, especially if the high type player is the first mover.

Table 8.2: Agreements on the equal split for $\alpha=45$

Type of first mover	$\alpha = 45$	Experience level			
		1	2	3	4
L	Total number of agreements	28	31	35	35
	Number of agreements on 50:50	16	24	26	29
	Number of immediate agreements on 50:50	4	6	7	8
H	Total number of agreements	19	15	23	24
	Number of agreements on 50:50	12	13	18	19
	Number of immediate agreements on 50:50	0	2	4	8

The proportion of agreements on the equal split is still fairly high, but the majority of these agreements is not reached immediately. The proportion of immediate equal split agreements increases from level to level, but only reaches 23% and 33%, respectively. The one-sided order test rejects the null hypothesis that there is no trend in the frequency of immediate equal split agreements in the four experience levels in favor of the alternative of an increasing trend, at a significance level of .01.

An immediately related question is the reaction of a player to the initial demand of the equal split. We have already investigated the number of first demands on the equal split made by the weak or the strong player that were accepted immediately. Now we pose the question of the type of the acceptor of these immediate equal split agreements.

Table 8.3: Number of immediate agreements on the equal split

α	Type of second mover	Experience level			
		1	2	3	4
30	L	2	11	15	19
	H	0	3	7	15
45	L	4	6	6	11
	H	0	2	5	5

Weak players are more likely to accept the equal split immediately than strong players.

THE GAMES WITH $\alpha > 50$

The agreement outcomes of two weak players in the plays of games with $\alpha > 50$ were highly asymmetric, and with a higher level of experience the asymmetry became even larger. In the following three figures (8.5 to 8.7) we shall look at the distribution of the initial demands of the weak players in the three parameter constellations for $\alpha > 50$. They show that the majority of the initial demands of the weak player is above α and moreover in the predicted outcome area for the strong player.

For $\alpha = 55$ the predicted area for the strong player's outcome is the interval from 60 to 80. The largest peak in the initial demands of highly experienced players is on 70, and the majority of the initial demands falls into this interval.

The same is true for $\alpha = 60$. Here, the predicted outcome area for the strong player is the interval from 65 to 80. The majority of the initial demands falls into the interval from 70 to 90. The distribution of the initial demands has two large peaks on 75 and on 80 in the upper part of the predicted outcome interval. Only one experienced player reveals his type directly by demanding less than α.

In the case of $\alpha = 70$, we saw a large number of agreement outcomes on the equal split, which was interpreted as the fear of the low type player of a break off by a possibly strong opponent. Consequently, directly revealing first demands on the equal split can be found. The majority of initial demands lies in the range from 70 to 90 with the major peak on 80, the middle of the predicted outcome interval for the strong player (from 75 to 85).

To summarize, the low type player, although facing a risk of break off by a strong opponent demands a large share of the coalition value as the first mover, in order to pretend to be the player with the high alternative. Generally, the demand is higher than α with the major peak in the middle of the predicted outcome range for the strong player. With a higher level of experience the demands tend to increase.

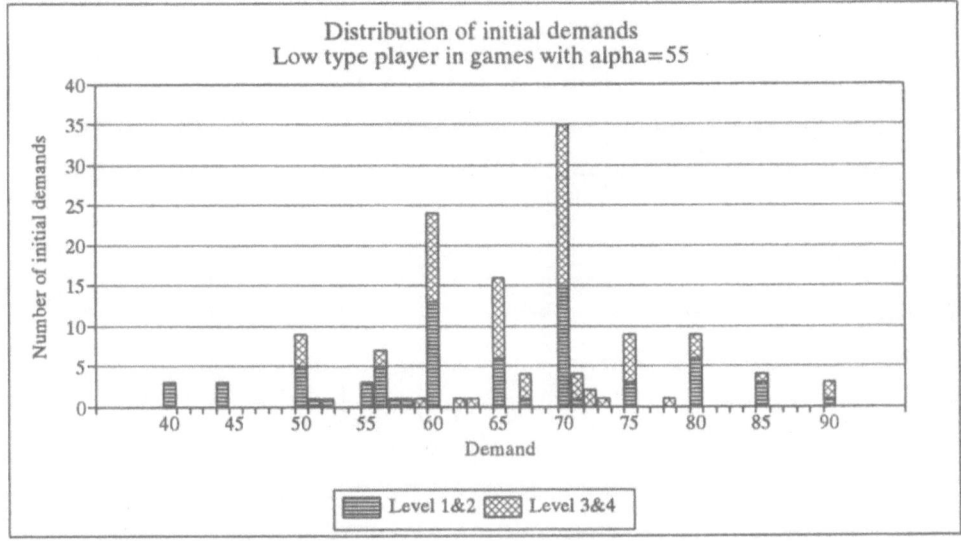

<u>Figure 8.5:</u>   Distribution of the initial demands of a low type player at the different
                experience levels for $\alpha=55$

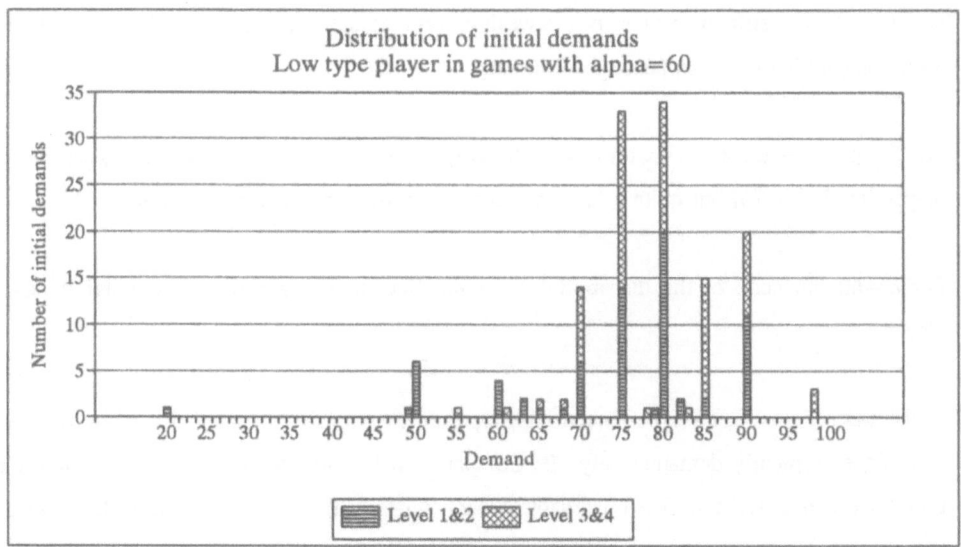

<u>Figure 8.6:</u>   Distribution of the initial demands of a low type player at the different
                experience levels for $\alpha=60$

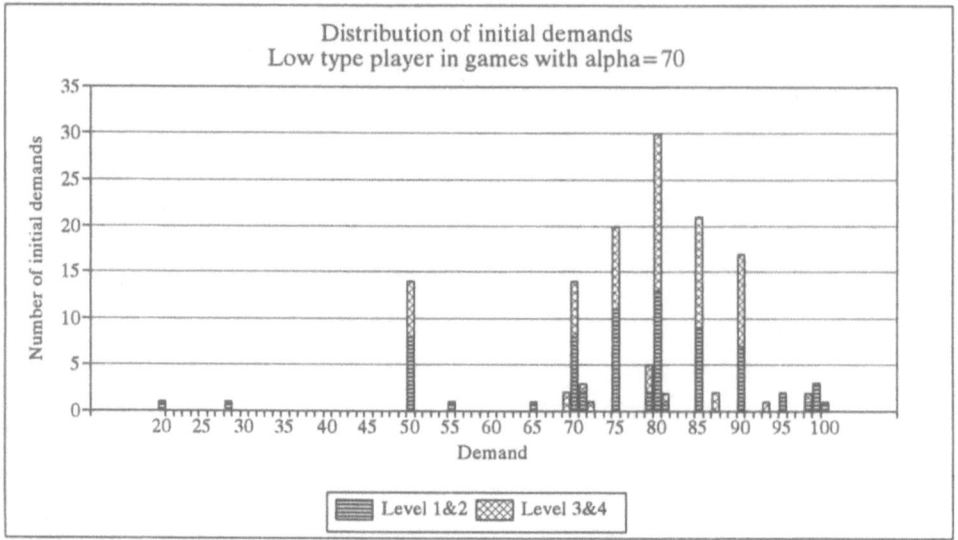

Figure 8.7:   Distribution of the initial demands of a low type player at the different
experience levels for $\alpha=70$

The distributions of the initial demands of the strong players in the plays of games with
$\alpha>50$ look very similar to those of the weak players. Usually, the major peak of the experienced players is one prominence level higher.

For the plays of the games with $\alpha=55$ the largest peak in the distribution for experienced
players is also at 70, but quite a large number of players also demand 75 or 80.

For $\alpha=60$ the peak of the distribution is at 80, like in the case of a first moving weak
player.

In the case of $\alpha=70$, an experienced high type player typically demands 90, while the low
type player typically demands only 80. Compared to the distribution of the weak player's
initial demands a right shift is observable, which supports the hypothesis that in the case of
$\alpha=70$ weak players are more cautious in their opening demand in order to avoid a possible
break off.

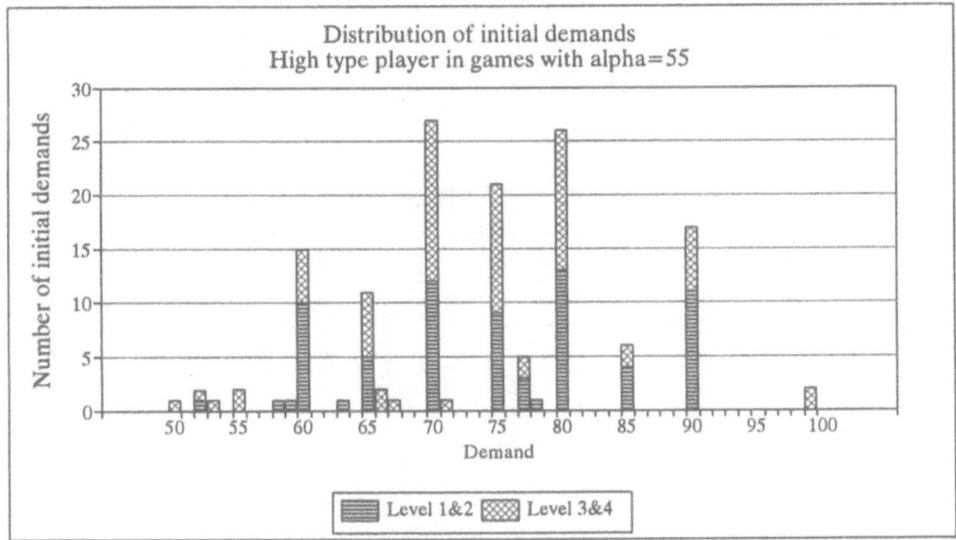

Figure 8.8:   Distribution of the initial demands of a high type player at the different experience levels for $\alpha=55$

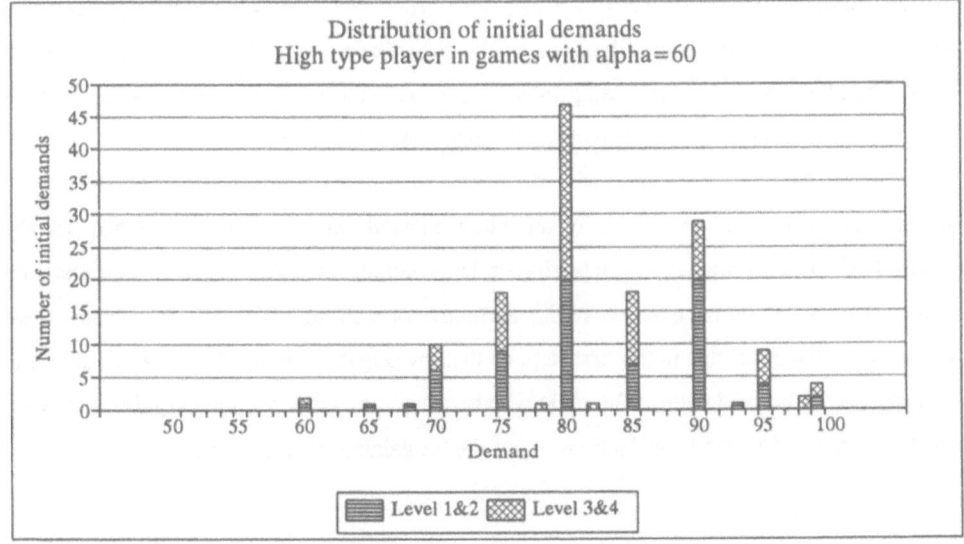

Figure 8.9:   Distribution of the initial demands of a high type player at the different experience levels for $\alpha=60$

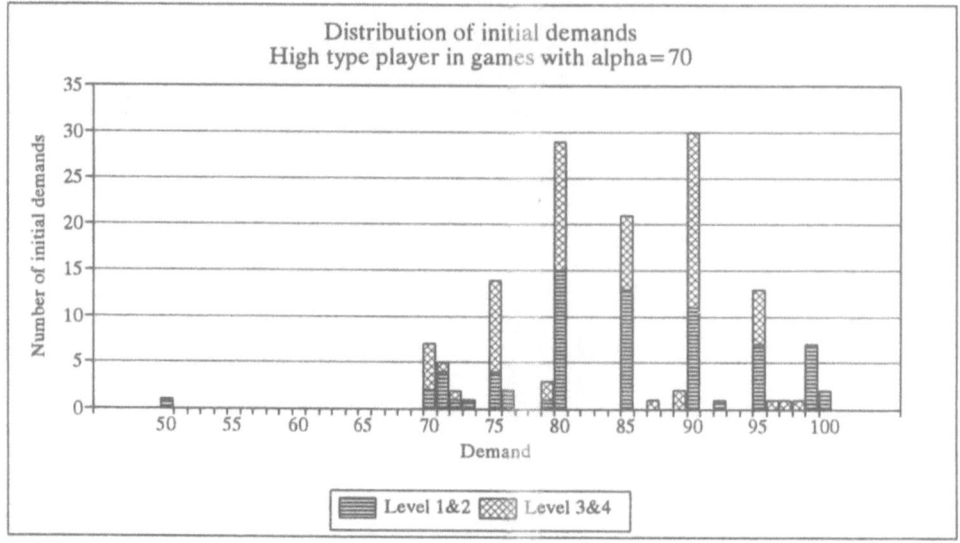

<u>Figure 8.10:</u>   Distribution of the initial demands of a high type player at the different
                        experience levels for $\alpha = 70$

For all parameter values of $\alpha > 50$ the one-sided order tests rejects the null hypothesis that
there is no trend in the initial demands of the experience levels at a significance level of
.107. The alternative of an increasing trend is favored in all three cases. Remember, that we
found a decrease in the initial demands with experience for $\alpha < 50$.

In the case of $\alpha < 50$ the success of an initial demand was discussed, especially for the
demand of 50. This was appropriate since a large number of plays ended in the first step
with an agreement. In the case where the alternative of a strong player is greater than 50 we
rarely observe agreements in the first step, so that investigations in the "success" of an initial
demand cannot be made here. The topic whether they may influence the bargaining process
shall be discussed in the investigations about the bargaining process in Chapter 9.

THE INITIAL DEMANDS IN DIFFERENT SITUATIONS

Although a weak player often imitates the strong player, the previous investigations indicate that the initial demand of the weak player is lower than the initial demand of the strong player. Actually, except for $\alpha=45$, the Wilcoxon matched-pairs signed-ranks test applied to independent subject group averages rejects the null hypothesis that there is no difference between the initial demand of a weak and the initial demand of a strong player at a significance level of .025 in favor of the alternative that the weak player's initial demand is smaller. For $\alpha=45$ this is only true for five of the six independent subject groups, which is not enough to reject the null hypothesis.

Up to now we only looked at the initial demand of the first mover. The question is whether the second mover responds to this demand with an equally high demand or whether he starts the concession process with his first demand. To respond with a lower demand might be dangerous since it is not obvious that the other player will also start to concede. On the other hand, it should be in the interest of the players to avoid discount steps. Except for $\alpha=30$, the Wilcoxon matched-pairs signed-ranks test applied to averages over independent subject groups rejects the null hypothesis that there is no difference between the initial demand of the first mover and the initial demand of the second mover at a significance level of .025. The alternative that the first mover has a higher initial demand is favored. For $\alpha=30$ this is only true for five of the six independent subject groups, which is not enough to reject the null hypothesis.

Nevertheless, it was not found that the first mover has a higher payoff than the second mover (see Section 6.2). The willingness to concede with the initial demand is not punished by a lower outcome. Apparently, the first mover is expected to make a high initial demand.

8.2   LENGTH OF BARGAINING

This section investigates the lengths of the bargaining until an agreement and a conflict, respectively is reached. Two tables that show the average bargaining length for plays that ended with an agreement and for those that ended with a break off for each gametype and for each experience level will be presented.

Table 8.4: Average bargaining length in agreement

α	Game-type	Experience level							
		1		2		3		4	
		#	avg	#	avg	#	avg	#	avg
30	(0, 0)	18	4.39	18	5.06	17	4.06	18	5.28
	(0,30)	15	8.87	16	10.00	15	5.80	16	5.63
	(30, 0)	14	9.29	17	6.71	17	7.76	16	4.69
	(30,30)	11	6.64	12	12.42	11	5.45	16	8.25
45	(0, 0)	18	7.44	18	4.22	18	3.61	18	4.22
	(0,45)	10	7.40	13	8.00	17	8.29	17	4.24
	(45, 0)	14	14.36	11	7.73	15	7.40	16	3.44
	(45,45)	5	7.20	4	11.75	8	5.75	8	6.25
55	(0, 0)	18	2.83	17	4.41	16	5.44	18	3.83
	(0,55)	9	6.89	13	4.38	9	12.11	6	8.67
	(55, 0)	5	4.80	8	5.13	8	9.00	8	15.13
	(55,55)	0	–	0	–	0	–	0	–
60	(0, 0)	17	3.47	16	11.38	17	10.59	18	13.11
	(0,60)	5	11.20	6	10.50	8	9.00	3	10.00
	(60, 0)	7	6.14	4	3.25	5	11.20	8	12.00
	(60,60)	0	–	1	7.00	0	–	1	12.00
70	(0, 0)	18	5.17	16	5.38	17	9.24	18	6.06
	(0,70)	4	9.00	4	9.50	2	3.00	3	5.67
	(70, 0)	2	10.00	5	7.4	3	7.33	3	8.00
	(70,70)	0	–	0	–	0	–	0	–

Table 8.5: Average bargaining length in conflict

α	Game-type	Experience level							
		1		2		3		4	
		#	avg	#	avg	#	avg	#	avg
30	(0, 0)	0	–	0	–	1	36.00	0	–
	(0,30)	3	11.67	2	10.00	3	26.33	2	29.00
	(30, 0)	4	9.00	1	18.00	1	13.00	2	23.50
	(30,30)	7	17.43	6	14.50	7	18.00	2	15.00
45	(0, 0)	0	–	0	–	0	–	0	–
	(0,45)	8	18.62	5	9.60	1	5.00	1	21.00
	(45, 0)	4	15.50	7	13.00	3	6.67	2	4.50
	(45,45)	13	8.15	14	9.50	10	11.20	10	9.20
55	(0, 0)	0	–	1	21.00	2	11.00	0	–
	(0,55)	9	4.78	5	5.20	9	9.00	12	10.25
	(55, 0)	13	8.15	10	5.40	10	7.40	10	9.30
	(55,55)	18	4.39	18	3.50	18	5.94	18	5.61
60	(0, 0)	1	6.00	2	17.50	1	75.00	0	–
	(0,60)	13	6.92	12	8.50	10	10.10	15	13.47
	(60, 0)	11	12.00	14	8.21	13	9.15	10	9.90
	(60,60)	18	5.06	17	5.35	18	5.39	17	5.59
70	(0, 0)	0	–	2	30.00	1	4.00	0	–
	(0, 70)	14	6.36	14	5.50	16	6.75	15	7.13
	(70, 0)	16	6.13	13	6.46	15	8.33	15	5.00
	(70,70)	18	3.89	18	4.50	18	4.50	18	4.11

For games ending in agreement, the average bargaining length has a tendency to decrease for plays of games with $\alpha < 50$ and a tendency to increase for $\alpha > 50$. This result should not surprise since the investigations in the agreement outcomes and the initial demands showed that for $\alpha < 50$ the players "learn" to agree on the equal split and this more often in the first step, if they gained experience.

For $\alpha > 50$ players increase their initial demand and weak players are able to reach more asymmetric agreements in plays with weak players. These "tougher" bargaining behavior leads to an increase of the bargaining length. But, the order test shows that these tendencies are only statistically significant for $\alpha = 45$ and $\alpha = 60$. For $\alpha = 45$ the one-sided order test favors the alternative of a decreasing trend in the bargaining length, and for $\alpha = 60$ it favors the alternative of an increasing trend in the bargaining length. Both at a significance level of .01. Hoggatt et al. (1978) also observed an increase in the length of bargaining for more experienced players. They explained this fact by the learning of players to avoid weak moves in a game which is comparable to our game with $\alpha = 45$.

In the high experience levels the average bargaining length for plays ending in conflict has a tendency to increase. It is remarkable that the average length of bargaining for plays that ended in conflict is about twice as long as the average length for plays that ended in agreement. This is approximately true for every gametype.

The bargaining between a weak and a strong player proceeds on average longer than the bargaining between two equally strong players. This is true for plays ending in agreement as well as for plays ending in conflict.

8.3   INCENTIVE COMPATIBILITY

The concept of incentive compatibility (see Myerson, 1979) states that "no player should expect any positive gains from being the only player to lie about his type when all others are planning to tell the truth". This means that for each type it should be more profitable to play according to the strategy of his type than to imitate the strategy of the other type. Since we are not able to observe the strategies of the players, the application of the concept of incentive compatibility to the experiment will be to compare the actually occurring average payoff of a type with the one which would result if he would imitate the other type.

For this purpose we shall evaluate the following variables:

A_L average (discounted) payoff of a low type player in agreement

A_H average (discounted) payoff of a high type player in agreement

p_L relative frequency of agreements of a low type player

p_H relative frequency of agreements of a high type player

Table 8.6 gives the values of these variables aggregated over all plays for each parameter of α. We do not distinguish between the first and the second mover, since we already saw that they are not distinguished in the average payoff and the average agreement frequency.

<u>Table 8.6</u>: Average agreement payoff and average frequency of agreement

α	Low type		High type	
	A_L	p_L	A_H	p_H
30	47.43	.93	47.20	.78
45	47.04	.89	47.97	.57
55	43.27	.71	60.95	.23
60	41.01	.63	64.32	.17
70	43.29	.57	72.35	.09

With the help of these variables the *incentive constraints* can be specified as follows:

$$p_L A_L \geq p_H A_H$$
$$p_H A_H + (1-p_H)\alpha \geq p_L A_L + (1-p_L)\alpha$$

The first inequality is the incentive constraint for the low type, and the second one is the incentive constraint for the high type.

The following table examines these constraints.

Table 8.7: Incentive constraints

α	Low type		High type	
	$p_L A_L$	$p_H A_H$	$p_H A_H + (1-p_H)\alpha$	$p_L A_L + (1-p_L)\alpha$
30	44.14	37.04	52.11	49.00
45	41.98	27.15	57.53	49.51
55	30.65	13.97	67.93	51.07
60	25.92	11.17	69.01	51.68
70	24.65	6.53	70.21	54.79

The incentive constraints are fulfilled for each parameter of α over all five treatments.

8.4   REVELATION BY THE WEAK PLAYER

The aim of this section is to investigate the revelation behavior of a weak player. Revelation of a weak player means that he demands for himself less than α. Under the assumption that a demand of less than α is not an error or an altruistic move by the high type player, such a demand reveals the low type player. In the following we shall study how this behavior affects the payoff of the revealer and the behavior of the opponent. In case that the opponent is also of type L, we shall examine whether he reveals his type, too. The main question asked will concern the profitability (in terms of average payoff) of a revelation.

For $\alpha < 50$ a revelation of the weak player occurred only in 5 of the 576 plays and this without exception in the first experience level. Accordingly, the point of interest are the plays of the games with $\alpha > 50$. If a low type player is playing a high type player, imitation bears always the risk of break off. The strategy of revealing the type and thereby having a potentially larger chance of an agreement (which is always more profitable than a break off) seems to be favorable.

For all plays of the games with $\alpha > 50$ where a low type player played a high type player the average payoff of a low type player that revealed and the average payoff of a non-revealing low type player was calculated. These data together with the number of revelations and non-revelations in all four experience levels is given in the following table. By non-revelation we mean, that a low type player never made a demand of less than α for himself. The table shows a surprisingly clear result. Except for $\alpha = 60$ at an experience level of 1, the average payoff of a revealer is higher than the one of a non-revealer. In the case of $\alpha = 55$ it is approximately twice as high, while the difference in the case of $\alpha = 70$ is even more strikingly. For $\alpha = 55$ nearly half of the players revealed and half of the players did not, whereas in the other two constellations of α there are approximately twice as many non-revealers as revealers.

Table 8.8: Average payoff by revelation and non-revelation of the weak player in plays of a weak player and a strong player

α	Experience level	Revealer		Non-Revealer	
		# plays	Avg payoff	# plays	Avg payoff
55	1	19	18.50	17	9.73
	2	17	24.55	19	13.92
	3	20	18.67	16	9.35
	4	17	13.70	19	9.50
60	1	10	7.61	26	9.69
	2	12	11.98	24	6.36
	3	12	18.10	24	3.56
	4	16	11.85	20	4.46
70	1	12	10.28	24	0.84
	2	14	14.24	22	0
	3	12	7.46	24	0.40
	4	14	6.06	22	1.14

We can conclude that, on average, the revelation of the true type was significantly more profitable than the non-revelation if the weak player plays the strong player.

However, a weak player does not know whether he faces a weak or a strong opponent. It might be possible that the revelation is so largely exploited by a weak opponent that it is no longer profitable. In fact, this turns out to be true. In order to see this we shall first look at the distribution of the agreement outcomes of a revealer in the case that he faces a low type opponent. The following three figure show these distributions for the three values of α.

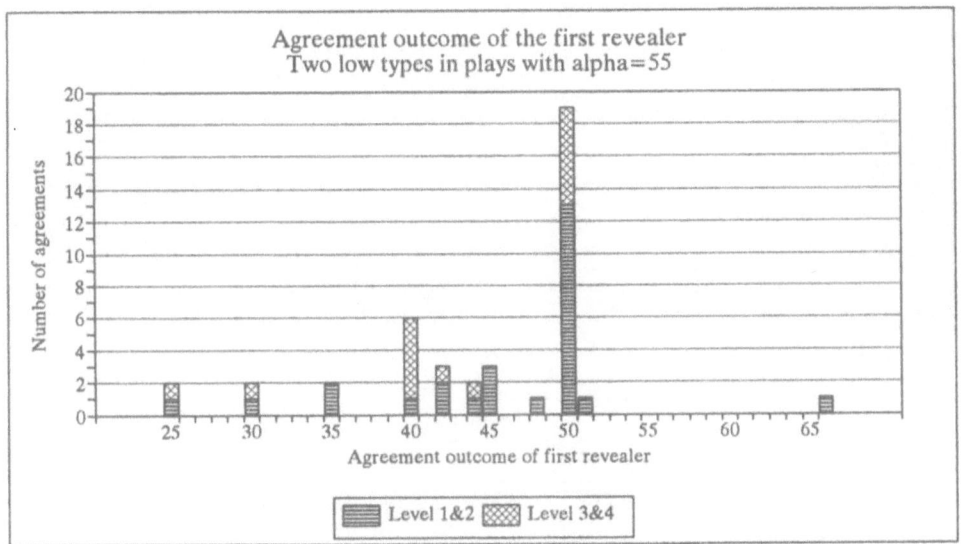

<u>Figure 8.11:</u>   Distribution of the agreement outcomes of a revealer playing a weak opponent for $\alpha=55$

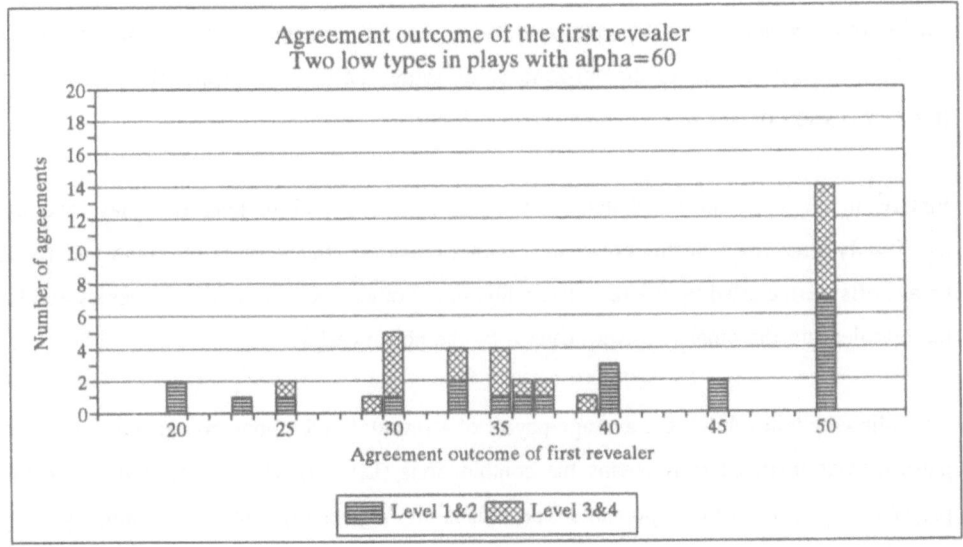

<u>Figure 8.12:</u>   Distribution of the agreement outcomes of a revealer playing a weak opponent for $\alpha=60$

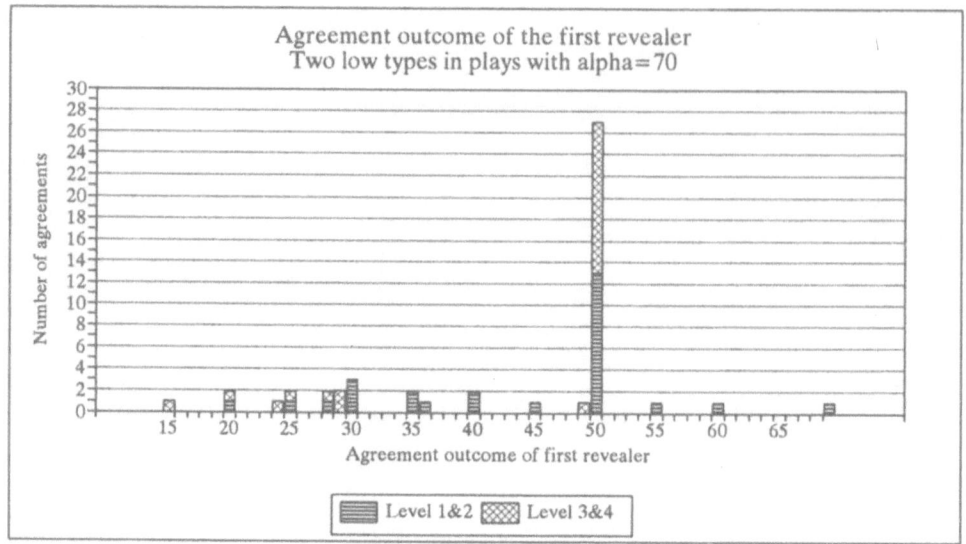

<u>Figure 8.13:</u>   Distribution of the agreement outcomes of a revealer playing a weak oppo-
                  nent for $\alpha=70$

For $\alpha=55$ and $\alpha=60$ the figures show a large asymmetry in the outcomes. Although, a large
number of agreements is on the equal split, the agreement outcome tends to be worse for the
revealer. For $\alpha=70$ most of the agreements are settled on the equal split after a revelation
(for a discussion of this phenomenon see also Section 6.2).

Finally, it has to be remarked that for the parameters 55 and 70 only two plays and for
$\alpha=60$ only three plays of this kind ended with a break off. In the cases of $\alpha=55$ and 70 all
breaks offs were caused by the revealer, while in the other case one break off was caused by
the revealer and the other two were caused by the non-revealer.

The following table shows the average payoff of a revealer and a non-revealer taken over all
plays he was involved (this means the combinations (L,L), (L,H), and (H,L)). Since the
player is not aware of the opponent's type, this is his "expectation" from revealing and non-
revealing, respectively.

Table 8.9: Average payoff by revelation and non-revelation of the weak player

α	Experience level	Revealer		Non-Revealer	
		# plays	Avg payoff	# plays	Avg payoff
55	1	33	30.85	39	32.46
	2	31	30.99	41	33.28
	3	31	24.84	41	31.33
	4	22	20.29	50	34.34
60	1	22	22.39	50	29.89
	2	25	20.83	47	26.01
	3	22	25.93	50	25.60
	4	27	19.93	45	29.98
70	1	28	27.85	44	24.82
	2	27	26.34	45	22.68
	3	24	22.96	48	23.37
	4	25	18.35	47	29.02

For $\alpha=55$ the non-revealer has the higher average payoff and it even increases with the level of experience. Similar results are obtained for $\alpha=60$, although for experience level 3 the average payoffs are nearly equal. This shows that overall a revelation was not profitable, although it is profitable in plays with strong players. For $\alpha=70$ a revealer has the higher average payoff in the low experience level and the converse is true for the high experience level. However, except for the fourth level, the differences are not drastic.

Similarly, an approximately equal amount of revealers and non-revealers can be found for $\alpha=55$ and for the other cases the ratio is again close to $\frac{1}{2}$. For $\alpha=55$ the one-sided order test rejects the null hypothesis of no trend in the revelation rate in favor of the alternative of a decreasing trend at a level of .107. For $\alpha=60$ and $\alpha=70$ H_0 cannot be rejected at a level of .2 (two-sided).

At first glance, these findings seem to be contradictory to the ones of the previous section. There we found that the data of the experiment is incentive compatible, which means that for no type it was profitable to imitate the other type. And now, we conclude that a non-revelation which could be seen as an imitation of the strong player is overall more profitable for the weak player. The difference between these two results is that for the verification of the incentive constraints, given the data of the experiment, the average payoff of all weak players (revealers and non-revealers) was calculated and compared to the average payoff of a strong player. In this section, however, we consider "strategies" of the weak and their payoff consequences.

The figures showed that a weak opponent took advantage of the revelation of the player by reaching no outcome worse than 50 unless he revealed himself. Now we want to examine whether a type H opponent also took advantage of a revealer. This will be done by comparing his average payoff from playing a revealer with the average payoff from playing a non-revealer.

Table 8.10:   Average payoff of the strong player playing a revealer and non-revealer, respectively

α	Experience level	Opponent is revealer		Opponent is non-revealer	
		# plays	Avg payoff	# plays	Avg payoff
55	1	19	52.19	17	56.97
	2	17	60.27	19	58.38
	3	20	57.07	16	53.08
	4	17	54.07	19	51.51
60	1	10	56.07	26	59.93
	2	12	56.06	24	56.98
	3	12	60.43	24	56.82
	4	16	61.25	20	57.54
70	1	12	56.98	24	67.44
	2	14	65.44	22	66.37
	3	12	69.77	24	65.22
	4	14	68.80	22	67.76

Table 8.10 shows that in the high experience levels the average payoffs of the strong player

playing a revealing opponent are higher than those playing a non-revealing opponent. This means that in the plays of games of a high type and a low type both sides take advantage of the revelation of the weak player. This phenomenon is explained by the higher agreement rate in case of a revelation. Table 8.11 shows the percentage of agreements of high types playing revealing and non-revealing low types.

Table 8.11:   Percentage of agreements in plays of games of the strong player playing a revealer and a non-revealer, respectively

α	Experience level	Percentage of agreements playing a...	
		revealer	non-revealer
55	1	47.36%	29.41%
	2	76.47%	42.11%
	3	65.00%	25.00%
	4	52.94%	26.32%
60	1	30.00%	34.62%
	2	50.00%	16.67%
	3	75.00%	16.67%
	4	50.00%	15.00%
70	1	41.67%	4.17%
	2	64.29%	0%
	3	33.33%	4.17%
	4	35.71%	4.54%

Except for the case of $\alpha=60$ and experience level 1, the agreement rates are strikingly higher playing a revealing opponent. These efficiency gains lead to higher payoffs for both players.

A comparison of the average agreement payoffs of the two players in case of revelation and non-revelation shows that the high type player reaches a higher payoff playing a revealer than a non-revealer. So, conditional on agreement a revealer is "exploited" by a strong opponent, but overall both players improve their payoffs by the revelation of the weak player.

The tremendous difference in the agreement rate for $\alpha=70$ explains the slight advantage of the revealer in the overall average payoff.

Another question one could pose is at which time of the bargaining the revelation occurs and how long it takes from then on to reach a result. Since these numbers do not systematically vary between the gametypes and the experience levels, we shall only give a rough description by naming the median of the averages of all sessions with $\alpha>50$: The revelation occurs in step 4 of the bargaining and the final result is reached after 5 additional steps.

BOTH WEAK PLAYERS REVEAL

In the plays of two low type players it occurred ten times (in all sessions with $\alpha>50$ together) that a player reveals after the opponent had revealed before. Only in one of these cases the first revealer reaches a lower outcome than the second revealer. In the other cases the first revealer has the higher outcome or the two players share the coalition value equally. By the revelation of the other player the first revealer looses his weak bargaining position and is able to reach the equal split or a higher outcome than the second revealer. The figures 8.11 to 8.13 show the cases in which the first revealer is able to achieve a higher outcome than the opponent.

The Binomial test is applied to the eight independent subject groups in which a revelation of both players occurs. We look at the two categories "the outcome of the first revealer is greater than or equal to the outcome of the second revealer" and "the outcome of the first revealer is smaller than the outcome of the second revealer". The null hypothesis that the observations are equally likely in both categories is rejected in favor of the alternative that the outcome of the first revealer is greater than or equal to the outcome of the second revealer (at a significance level of .01, one-sided).

8.5   ALTRUISM

Bolton (1991) states that "bargainers measure what they receive by both an absolute and a relative yardstick". He observes a large number of disadvantageous counteroffers which he explains by the interest of players in the relative payoffs. If they are seen as unfair, bargainers are willing to loose money in order to achieve more equal allocations. The disadvantageous counteroffers he observed were of the form that the proposer increased the number of "chips" for himself, but nevertheless lost money due to discounting. This means that these players are concerned about fairness for themselves and not for the opponent.

Because of the low discount factor of 1% and a smallest money unit of 1 in our experiment this type of disadvantageous counteroffers could not occur. Nevertheless, it happened five times that a proposer repeated the allocation of the opponent (which was then accepted immediately by the opponent). This means, instead of accepting, the proposer repeated the allocation and thereby "wasted" one discount step. These five plays were played by different players and three of them in the first experience level. I tend to view them as mistakes by the subjects.

There is an experiment by Frohlich and Oppenheimer (1984) which systematically studies the occurrence of non self-interested behavior like altruism. *Altruism* is understood as the player's renunciation of payoff in order to increase the opponent's payoff. The experiment lasted for one week and at each day each subject had to perform several binary choices. Among them there were three pairs which attempted to investigate the phenomenon of altruism. In each pair one choice was the payoff configuration (8,7), where the first tuple component gives the payoff of the player and the second component is the payoff of the opponent. The second pair always was the "altruistic choice" with (7,14), (5,14), and (3,14), respectively. By choosing one pair the subject determined the payoffs of this game. No confirmation by the opponent was needed. Frohlich and Oppenheimer conducted this experiment in Canada as well as in the US. In Canada in 20.6% of the cases an altruistic option was chosen, and the occurrence of this phenomenon was surprisingly stable over the whole week under observation. In the United States an altruistic option was chosen in 38% of all cases, also with a high stability in the whole week.

The occurrence of altruistic behavior is surprisingly high. The authors suggest that the US result could be a consequence of the subject pool. The US students mostly were "liberal art" students, while the Canadian students mostly were business students.

An important difference between the payoff structure of the underlying experiment and the study by Frohlich and Oppenheimer is that in our experiment the payoff sum is always constant. In the experiment by Frohlich and Oppenheimer a player could double the opponent's payoff by decreasing his payoff by 1. This means that the joint profit was maximized by the altruistic choice.

Forsythe, Kennan and Sopher (1991) find in their experiment that altruism is not a typical feature, but has a "substantial influence on the outcome".

However, the occurrence of altruism was found in previous experiments and it should be examined whether altruism is of "serious" importance in our experiment. We shall analyze for our data how often it occurred that a player demanded for himself less than the other player offered to him, or accepted less than his alternative value.

It happened in 10 plays of a strong and a weak player with $\alpha=55$ that an equal split agreement was reached. Seven times, the equal split was proposed by the weak player and accepted by the strong player, and in three cases the final agreement was proposed by the strong player. One player was involved three times as the strong player in these games. After he experienced twice that the weak opponent made the equal split proposal and he accepted, he himself proposed in the third play the equal split as a strong player. One player was involved twice, the first time as the proposer and the second time as the acceptor of the equal split. The remaining five plays were played by different subjects. For $\alpha=60$ it happened four times that a strong player agreed on a split below his alternative value. The outcomes were 7, 20, 21, and 30, respectively. Since in all cases the strong player was the acceptor and the strong player always receives a minor share, these results might be seen as "acceptances by mistake". In plays of $\alpha=70$, two (different) strong players agreed to 50:50 as acceptors.

Concerning the disadvantageous counteroffers, we, contrary to the investigation by Bolton, look at counteroffers which yielded a smaller number of points to the proposer than the opponent offered to him. This means that the discounted as well as the non-discounted (point) value was smaller.

However, this behavior might be due to altruistic behavior or due to a typing error of the subject. Since this is unknown to the experimenter all cases will be listed in the following table, so that the reader has the possibility to decide.

Table 8.12:   Listing of plays is which a player proposes less for himself than offered by the opponent (marked with a "•"). These plays were all played by different subjects.

Gametype		Experience level	Proposer	Proposal for player 1	Proposal for player 2	Accepted?
30	(0,0)	1	1	40	60	
			2	49	51 •	
			1	51	49	Acc by 2
55	(0,0)	3	1	71	29	
			2	65	35	
			1	60 •	40	Acc by 2
55	(0,0)	1	1	40	60	
			2	66	34	
			1	55 •	45	
			2	66	34 •	Acc by 1
55	(0,0)	2	1	56	44	
			2	40	60	
			1	42	58	
			2	50	50 •	
			1	58	42	
			2	45	55	
			1	50	50	Acc by 2
55	(0,0)	1	1	40	60	
			2	50	50 •	Acc by 1

To summarize, altruism seems to have a negligible influence on bargaining.

CHAPTER 9. THE BARGAINING PROCESS

9.1   MODELS OF THE BARGAINING PROCESS

Already the early approaches to the bargaining problem were concerned with a formalization of the bargaining process. Harsanyi (1956) gave an interpretation of Zeuthen's (1930) bargaining theory in the light of game theory. In this interpretation the bargaining process is driven by the concessions of the players, which emerge as repeated improvements of a product of utility differences. A player will make a concession if for his proposal this product is lower than the product resulting from the opponent's proposal. In case of equality of the utility products Zeuthen assumes that the player will concede. The magnitude of the concession is not specified explicitly. The difficulty of this approach is that the utility functions of the players have to be known. Hence, a test of this theory (see for example Tietz and Weber, 1972) requires to make assumptions on these functions.

The problem that not all components of the bargaining theory are known is widespread also in the further attempts to explain the bargaining process. There is a variety of approaches that intend to explain the bargaining process by the aspirations of the bargainers. Since aspiration levels are not observable from the data there are potentially two different ways to reveal them. The first way is to force the subjects to fill in a "planning questionnaire", which contains the five major *aspiration levels* according to Tietz (1978). These five aspiration levels are the *planned* (P), the *attainable* (AT), the *acceptable* (AC), the *threat* (T) and the *break-off* (L). At the threat level a player threatens the break off and at the break off level he actually breaks off. Such a planning questionnaire was used in Tietz and Weber (1972), Bartos, Tietz and McLean (1983) and Tietz et al. (1978) to reveal the subjects' aspiration levels.

The second way of being knowledgeable on the aspiration levels of the subjects is to induce them. Crott, Müller and Hamel (1978) conducted a repeated two-person bargaining experiment in which a selected group of subjects played the game three times, always playing inexperienced opponents. They induced an aspiration level in the subjects by announcing a doubling of the gain if the subject was able to reach at least a certain payoff. Crott et al.

found that subjects with a high aspiration level had higher gain expectations, made higher initial demands, and were able to achieve higher gains. Furthermore, an increase of these values was observable with an increase of experience. With experience also the bargaining length increased.

However, these kinds of analysis of the bargaining process need to know the aspiration levels of the subjects and therefore their applicability is only restricted.

There are other approaches which refrained from questioning planning levels but estimated the bargaining model from the data. Bartos (1974) hypothesizes that a player's demand is a linear function of the last demand of the proposer and the last demand of the opponent:

$$D_t = a\,D_{t-2} + b\,D_{t-1} + c\,, \qquad 0 \leq a \leq 1,\ -1 \leq b \leq 1,$$

where D_t and D_{t-2} are two consecutive demands of the negotiator and D_{t-1} is the intervening demand of the opponent.

Bartos, Tietz and McLean (1983) examined this model with experimental data and concluded that the opponent's behavior has no impact on the negotiator, hence the equation can be simplified by assuming that $b=0$ as

$$D_t = a\,D_{t-2} + c.$$

The result was shown by the evaluation of the reaction matrix with help of the χ^2 - test.

Bartos (1978) investigates the concession behavior of players in a two-person bargaining game. He found that the subjects made their largest concessions when the opponent made low offers. This means that they take advantage of a weak opponent (making large concessions), and are pressured by a tough opponent to make large concessions themselves. An a posteriori model of "fair concessions" failed in the light of the data of the experiment.

Cross (1965) introduces a bargaining model where the players' demands depend upon their beliefs on the opponent's concession rate. During the bargaining process the players correct their beliefs according to the observed concession behavior of the opponent.

A very interesting study has been conducted by Yukl (1974). In his experiments the subjects bargained with a programmed opponent. The subjects were told that the opponent is another subject from the recruited group. The special procedure served for the controlled manipulation of concession attitudes, such as initial offer, final offer, concession frequency, concession speed, and concession magnitude.

The first experiment examined two levels of concession magnitudes and three levels of initial offers. All six combinations were played. All subjects were (privately) told that they are sellers and they should try to sell a used car which is worth $2,500 to the buyer at the highest possible price. The excess was their profit. According to their profit the subjects earned credit points in their "industrial psychology course" and moreover the best buyer and the best seller was awarded with $5. The first offer was always made by the (programmed) buyer. The bargaining process was finite with a randomly determined end. The subjects knew that each party would be able to make between eight and ten offers, but that the exact number and the last proposer were determined by chance. With each proposal they had to concede at least $5. The *target point* of a subject (best expected outcome) and the *resistance point* (lowest acceptable price) were asked by a questionnaire prior to each subject's offer.

Yukl found that the opponent's initial offer had a significant effect on the subject's initial offer and the subject's final offer. There was no significant effect of the opponent's initial offer on the subjects concession magnitude. The subject's final offer was more favorable when the opponent made only small concessions than when the opponent made large concessions. Subject's concession magnitude was not affected by the opponent's concession magnitude. The harder the opponent's initial offer and the smaller his concessions, the lower was the subject's target point. The harder the opponent's initial offer and the smaller his concession, the lower was the subject's resistance point. These observations are consistent with the aspiration level hypothesis (Siegel and Fouraker, 1960).

A second experiment was designed to examine whether the subject is affected by the opponent's concession magnitude if the opponent does not concede in each step. Therefore, two levels of concession (frequent and infrequent) and two levels of concession magnitude were tested. The further features of the setup were like in the first experiment with the exceptions that all subjects were now buyers and the maximal bargaining length was 13 to 14 steps. The subjects were not forced to make a concession in two consecutive offers, but they were also not allowed to increase their demands.

Yukl found that the subjects conceded more frequently if the opponent conceded more frequently. From his findings he concludes that "the aspiration level hypothesis applied to the effect of opponent concession magnitude only when the opponent made frequent concessions". And, "a hard initial offer is an effective strategy for lowering a bargainer's aspiration level and obtaining a favorable final offer from him when he does not have prior information about the opponent's payoffs. A strategy of small opponent concessions is superior to a large concession strategy only when opponent concessions are frequent and the opponent's offers allow the bargainer to obtain positive payoffs. In addition, a small concession strategy may not be effective if there is no time pressure or the bargainer has complete information about opponent payoff." "Concession frequency did not appear to be an important concession parameter, except as a moderator of the effects of concession magnitude and possibly of initial offer."

9.2   THE CONCESSION BEHAVIOR IN THE BARGAINING PROCESS

A *concession* is the difference between two successive demands of a player. In case of a decrease of the demands the concession is positive, otherwise it is negative. For each of the six independent subject groups the concessions of all plays that ended in agreement are evaluated. The evaluation consists of the verification of claims about the concession behavior. For each parameter value of α a Wilcoxon matched-pairs signed-ranks test is conducted with the six independent subject groups to examine whether the number of cases in which the claim is true is significantly greater than the number of cases in which it is false. The cases in which the null hypothesis that there is no difference in these numbers can be rejected in favor of the alternative that the claim is true (at a significance level .025, one-sided) are marked with a "•" in table 9.2, at the end of this section. Moreover, the table gives the percentage of cases in which the claim is actually correct. The Wilcoxon matched-pairs signed-ranks test only detects a "statistical correctness" which does not imply a correctness in each case.

<u>Claim 1:</u>   *A player always makes positive concessions.*

The first claim is concerned with the direction of the concession. Independent of the past decisions of the opponent a player never increases his demand. It may be that he repeats his last demand, but he never increases it.

<u>Claim 2:</u>   *If the opponent repeated his demand, the player makes a concession.*

A demand repetition by the opponent is not followed by a demand repetition by the player.

<u>Claim 3:</u>   *The concession process of a player is decreasing.*

The concessions remain equal or shrink from bargaining step to bargaining step. This means that the players start with large concessions, but they shrink in the course of the bargaining. Table 9.1 is designed to give an insight in the magnitude of the average concession in a play that ends in agreement. It distinguishes the average of the strictly positive and the average of the strictly negative concessions.

Table 9.1: Average concession in plays ending in agreement. The numbers in brackets are the averages, where concessions from (and to) 99 or 100 are excluded. The average concessions in a monotonically decreasing proposal process are not explicitly listed in the table, since they nearly coincide with the numbers in brackets.

α	Avg of strictly positive concessions	Avg of strictly negative concessions
30	7.07 (6.00)	-8.53 (-5.35)
45	9.12 (7.75)	-9.65 (-5.86)
55	7.79 (7.74)	-6.44 (-6.31)
60	8.37 (8.02)	-4.45 (-3.78)
70	10.86 (10.69)	-7.65 (-6.12)

The next claims show that the concession process is influenced by the actual level of the demand. The players seem to have certain values which are approached more "carefully" than others. This means that the concessions shrink overproportionally as they approach these values. All numbers, which are divisible by 10 without remainder are values "with a high level of resistance".

Claim 4:   *If the previous demand of the player exceeded a number divisible by 10 (without remainder) by $1 \leq r \leq 4$ the concession from the previous demand to the actual demand is at most r.*

Suppose, the previous demand was 63, which exceeds a number divisible by 10 by $r=3$. The claim states that the actual demand is not smaller than 60. It may be that the player increases his demand, repeats it or makes a small concession of at most 3. A comparison with the previous table shows that these concessions are extremely below the average, for every parameter value of α. Even if one looks at the concessions which are made without an increase in the demand before.

The players are cautious in approaching these values and reluctant to jump over them in one bargaining step. In the terminology of the aspiration levels (see Section 9.1) these values can be seen as the planned aspiration levels of the players. The players approach them as planned outcomes, but if they see that they cannot fulfill an aspiration level they abandon it and switch to a new one.

For $\alpha=30$, 45, and 55 an even stronger version of claim 4 can be stated.

Claim 5:   *If the previous demand of the player exceeded a number divisible by 10 (without remainder) by $1 \leq r \leq 3$ the concession from the previous demand to the actual demand is at most r.*

One would expect that it is possible to further strengthen the result by considering only a subset of the numbers divisible by 10, for example 50, 60, and 70. But, by any restriction to a subset of the numbers which are divisible by 10, it is not possible to further strengthen the claims 4 and 5.

It can be supposed that prominent numbers divisible by 5 also play an important role as points of stronger resistance. But, claim 6 is only partially true (only for $\alpha=30$, 45, and 55).

Claim 6:   *If the previous demand of the player exceeded a number divisible by 5 (without remainder) by $1 \leq r \leq 3$ the concession from the previous demand to the actual demand is at most r.*

Besides the numbers divisible by 10 other focal points can serve as aspiration levels of the players. Especially, it will be tested whether the value 50 (equal split) and the split of the difference are approached in the way described above.

Claim 7:   *If the previous demand of the player exceeded 50 by $1 \leq r \leq 2$ the concession from the previous demand to the actual demand is at most r.*

There is a tremendous reluctance to lower the demand below the equal split. Even if the previous demand was 51 or 52, the actual concession will not drop the demand below 50. This very strong result is valid for all parameters of α, besides for $\alpha=60$. For $\alpha<50$ the major influence of the equal split was demonstrated by the agreement outcomes. For $\alpha=70$, the equal split is no longer a natural division for two strong players, but we observed a large number of equal split agreements for two low type players. The fear of break off makes the equal split a focal point and leads to such a strong result.

The analysis of the agreement outcomes revealed the importance of the equal split of the surplus as a distribution scheme. The following claim examines whether these values also serve as aspiration levels of the players, this means whether they are approached very slowly. Especially, we look at the equal split of the difference in addition to α, rounded to the next higher prominent number. These are the values 65 for $\alpha=30$, 75 for $\alpha=45$, 80 for $\alpha=55$, 80 for $\alpha=60$, and 85 for $\alpha=70$.

Claim 8: *If the previous demand of the player exceeded the equal split of the surplus in addition to α by $1 \leq r \leq 4$ the concession from the previous demand to the actual demand is at most r.*

Only for the high alternative values $\alpha=60$ and $\alpha=70$ this claim is of importance, which is in accord with the findings on the agreement outcomes.

Table 9.2 shows the percentages of correctness of each claim for each parameter value of α. Furthermore, a "•" indicates whether the result is significant by the Wilcoxon matched-pairs signed-ranks test (at a significance level of .025, one-sided).

Table 9.2: Percentage and significance (marked by a "•") of correctness of the claims

Claim	$\alpha = 30$	$\alpha = 45$	$\alpha = 55$	$\alpha = 60$	$\alpha = 70$
1	81.31% •	80.35% •	81.13% •	78.29% •	81.84% •
2	62.80% •	64.54% •	85.53% •	69.19% •	63.73% •
3	67.81% •	68.46% •	67.47% •	64.16% •	63.00% •
4	92.05% •	95.28% •	81.54% •	82.54% •	78.57% •
5	91.08% •	95.57% •	77.78% •	78.57%	74.42%
6	82.41% •	86.45% •	74.77% •	77.33%	73.56%
7	99.18% •	100% •	100% •	66.67%	100% •
8	52.73%	45.71%	28.30%	87.80% •	83.33% •

9.3   THE BARGAINING PROCESS IN GAMES WITH $\alpha < 50$

The analysis of the bargaining outcomes for plays of games with $\alpha < 50$ showed that the striking majority of the agreement outcomes was on the equal split, and on the high experience level the average deviation from the equal split was less than 1. Nevertheless, a bargaining process with an average length of roughly five steps was needed to reach these results. In this section we shall study the bargaining processes that ended in agreement. The circumstances of break offs were studied in a previous chapter (see Chapter 7).

CORRELATION OF BARGAINING CHARACTERISTICS

A first insight into the bargaining process will be given by a correlation analysis of the player's bargaining characteristics, similar to the investigations by Yukl (1974) (see Section 9.1). For each play the following features are evaluated for both bargaining partners: the *number of concessions*, the *concession frequency* (the number of concessions relative to the number of demands by the player), the *number of large concessions* (this are concessions that exceed 5), and the *concession rate* (the total concession in proportion to the initial demand). For each parameter value of α a Spearman rank correlation test was conducted and table 9.3 shows whether the null hypothesis that the considered characteristics are not correlated could be rejected at a level of .05 (one-sided) for the following pairs of characteristics.

Table 9.3: Positive correlation of bargaining characteristics (marked with a "•")

Characteristic 1	Characteristic 2	$\alpha=30$	$\alpha=45$
Initial demand	Number of concessions	•	
Initial demand	Concession frequency		•
Initial demand	Number of large concessions	•	•
Initial demand	Concession rate	•	•
Concession frequency 1. mover	Concession frequency 2. mover	•	

In all cases of a rejection of the null hypothesis (marked with a "•" in the table) the alternative of a positive correlation was favored.

For both parameter values of α it is true, that the higher the initial demand of a player is, the higher is his concession rate. This result does not surprise since the agreement outcome is always close to 50. It is more interesting that the players with the high initial demands also make a large number of large concessions. In order to avoid discount steps the concessions are high. For $\alpha = 45$ the concessions of a player with a high initial demand are also highly frequent. The correlation of the concession frequency of the first mover and the second mover is the only characteristic in which the players influence each other significantly.

THE SHAPE OF THE BARGAINING PROCESS

Although the agreement outcome of the bargaining was quite uniform for plays of games with $\alpha < 50$, there is a considerably large number of "long" bargaining processes. Among the 576 plays of games with $\alpha < 50$ there are 457 plays which ended in agreement (79%). In 119 (26%) of these plays the first demand was immediately accepted, 170 (37%) plays lasted between two and six steps, and 168 (37%) plays had a duration of more than seven steps. In a bargaining ending in agreement after seven steps the first mover makes four demands and the second mover proposes three times and accepts once. This means every player has four moves.

We shall address the question of the shape of the bargaining process. A first look at the bargaining processes hints at a common feature of most processes. The players start with high initial demands, make large concessions in the beginning, and, as the they come close to the equal split, the concessions "inch" to that goal. This observation is in accord with the findings in Section 9.2. A typical example for such a proposal process is: 70, 60, 55, 53, 52, 50. The shape of this process is close to the shape of an exponential process with a negative slope. Accordingly, an appropriate approximation of the bargaining process could be by an exponential function.

But, some of the bargaining processes do not have a "pure" form as described above. The instrument of increasing the demand is sometimes used by the players. Most demand increases occur either in demands of 99 and 100, or in increases of one or two points. These increases have to be seen as manipulative moves, since they are always followed by a demand close to the previous one. This means that they do not increase the long-term demand level.

However, for the approximation of the bargaining process leading to an agreement by an exponential function we shall look at all plays of more than seven steps and among them at all demand processes without increases in the demands. The first restriction was chosen since it does not seem to be adequate to estimate processes unless each player has at least four moves. Furthermore, it is not possible to find an appropriate fit of a demand process including threats (and the meaning of such an approach is questionable). One way to overcome this problem is to ignore these threats. But then one has to decide whether to assume different demands for these steps or to shorten the bargaining process. Since there seems to be no natural solution for this problem, I decided to concentrate on the bargaining processes without demand increases.

For each demand process without demand increases that occurred in a play ending in agreement after seven or more steps a least-squares approximation with a function ae^{bx} was determined. The result is strong. In 72% of all estimations a coefficient of determination (R^2) of more than .8 and in 48% of all cases a coefficient of determination of more than .9 was reached.

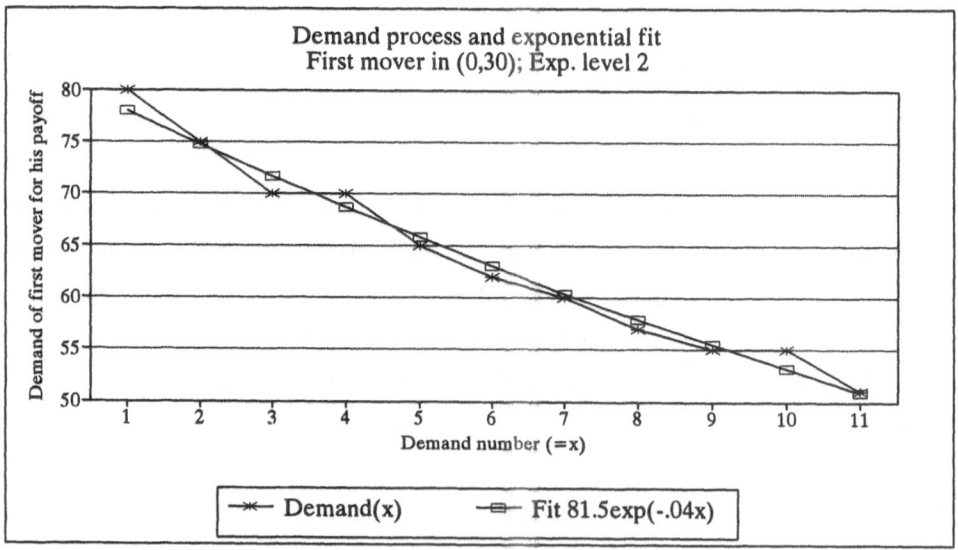

<u>Figure 9.1:</u>   Exponential fit of a demand process

The figure shows an example of such an exponential fit. The least squares fit of the demand process of a first mover in a play of a (0,30) game is reached by the function $81.5e^{-.04x}$ with a coefficient of determination of .9829.

Since in most of the cases, as in the above example, the curvature of the exponential function is not so strong, an alternative estimation of a linear approximation of the form ax+b was also determined. But, the goodness of these fits is worse than in the exponential case.

The intuitive idea of an approximation of the demand process by an exponential function is supported by the strong results of the least-squares fits. One possible interpretation of this result can be given by the aspiration level approach by Tietz (1978). The fair solution as the planned aspiration level is approached by the bargaining process. Since the bargainer hesitates to demand below this value, the concessions shrink as they approach it. The investigations in Section 9.2 showed that in nearly 100% of all cases a demand of 51 or 52 was followed by a demand not lower than 50, although the average concession is about 8 and 9, respectively.

This finding together with the conformity of the agreement outcomes for $\alpha < 50$ might suggest to choose an approximation by an exponential process of the form $y=50+ae^{bx}$, which means an approximation of $y-50$ by the exponential process ae^{bx}. An exponential approximation of this kind is transformed into a linear approximation by taking the logarithms of the y values. Since we observed demands (y values) of less or equal 50, the difference $y-50$ can be lower or equal to zero, which destroys the possibility of the reduction to a linear problem by taking the logarithms. An approximation which refrains from the transformation to logarithms causes serious numerical problems.

The approximation of the demand process by an exponential function neglects the influence of the opponent on the player's demands. Of course, the opponent's demands have an influence on the player's demands. But, for the plays of games with $\alpha < 50$ the structure of the bargaining processes is so much alike that the approximation can neglect this influence and concentrate on the concession process of each player. Accordingly, the previous investigations revealed that the initial demands as well as the outcomes of plays with experienced players do not show much diversity.

Nevertheless, the question that arises is how the estimated parameters a and b look like and how they are tied to the underlying demand process. For all approximations with a coefficient of determination greater than .9, the parameter a is greater than the initial demand. On average a exceeds the initial demand by 5.6. The exponent b is always negative and the absolute value is smaller than .1. On average over all estimations with a coefficient of determination greater than .9 it is $-.08$.

It shall be illustrated how exponential processes of this form look like and how parameter changes influence their shape. For the sake of simplicity this will be done by studying different types of processes graphically, in absence of a concise mathematical analysis.

The following two figures show exponential processes where one parameter is kept fix and the other one is varied. The base process in both cases is $76e^{-.08x}$. Assuming an initial demand of 70, this is the process with the average parameters of the estimations.

The first figure varies the parameter b with $b=-.06$ and $b=-.10$. The starting points (estimated initial demands) of these processes are very close, but by the different slope the processes grow apart. Thus, the cutting point of the equal split is reached in 5 steps for $b=-.10$, 6 steps for $b=-.08$, and 8 steps for $b=-.06$.

The second figure varies the parameter a with $a=72$ and $a=80$. Still assuming an initial demand of 70 this means varying the excess by 2 and 10. The resulting exponential processes are much more parallel, such that the cutting point of the equal split is in step 5 for $b=72$, and 6 for $b=76$ and $b=80$.

The major influence on the process is given by the exponent b, which determines the concession speed. The parameter b is always negative. The higher the absolute value of b is, the higher is the concession speed of the exponential process. The parameter a mostly determines the height of the process.

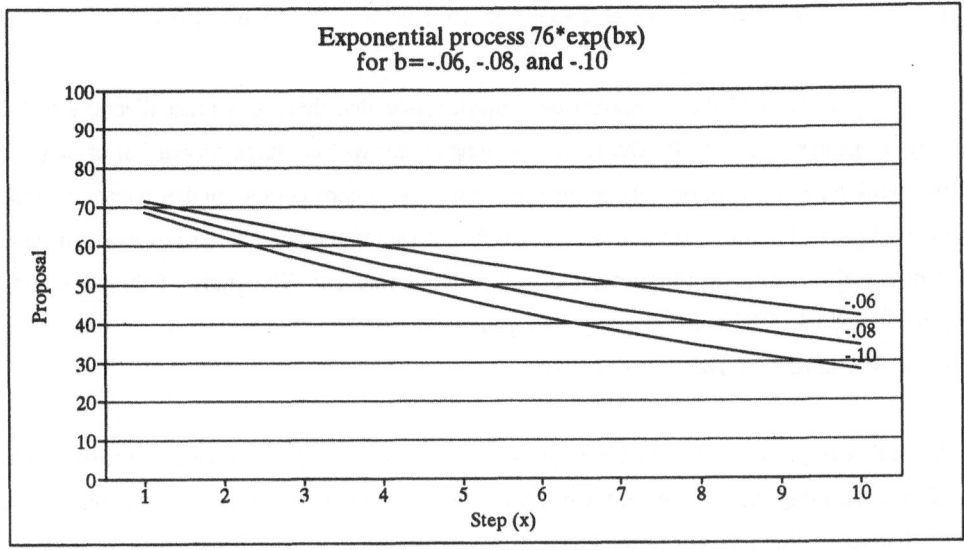

Figure 9.2:   Exponential process for different values of b

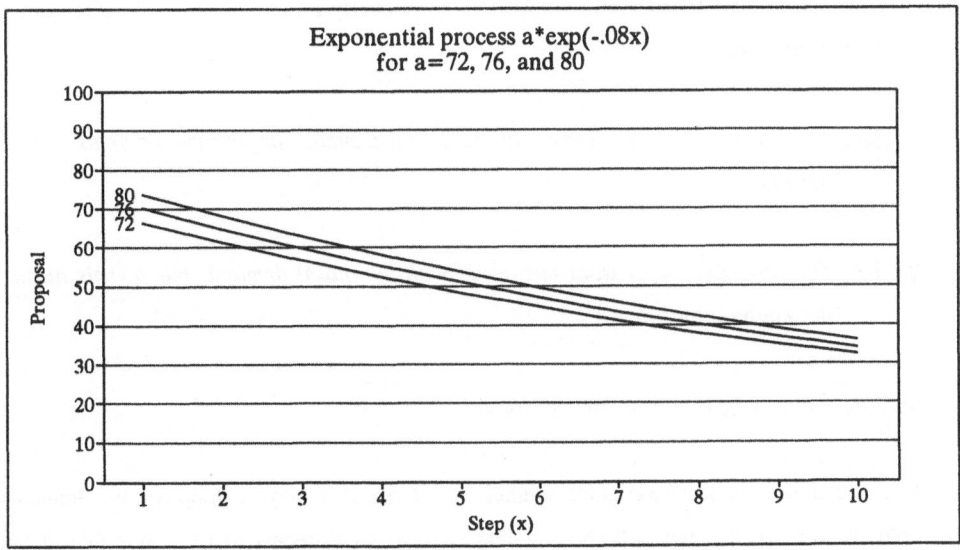

Figure 9.3:   Exponential process for different values of a

9.4   THE BARGAINING PROCESS IN GAMES WITH $\alpha > 50$

The analysis of the agreement outcomes showed that there is a great diversity in the plays of games with $\alpha > 50$. The most intriguing aspect was the large proportion of plays of two weak players where one player managed to receive a considerably higher payoff than the other player. The aim of the following analysis is to study the negotiation processes of plays with $\alpha > 50$ in order to illuminate the way to the agreement. The study of the bargaining process will be restricted to the plays that ended in agreement. The circumstances that led to break offs were analyzed in Chapter 7.

According to the revelation of the weak player we shall distinguish three exhaustive shapes of the bargaining process. Since the distinction is due to the behavior of the weak player, we shall speak of three shapes of the behavior of the weak player. The first shape describes the immediate revealer who's initial demand is lower than α. The second shape captures a weak player who did not reveal with his initial demand but demands less than α during the negotiation process. Finally, we shall look at a weak player who does not reveal during the whole negotiations. In case this player plays a strong opponent he reveals by accepting a demand which leaves less than α to him and an individually rational amount of more than α to the strong opponent.

Shape L1: *The low type player reveals with his initial demand (as first mover or as second mover)*

Shape L2: *The low type player does not reveal with his initial demand, but reveals during the negotiation process*

Shape L3: *The low type player does not reveal*

A very interesting phenomenon is the *sudden acceptance* of a player. Suppose, the demands of both players are on such a high level that an agreement seems to be impossible at the current stage. But, suddenly, one player accepts (the high) demand of the opponent. We shall define that a sudden acceptance occurs if the accepted share is more than ten units below the minimal value the player demanded for himself during the whole negotiation process. The value ten is chosen arbitrary, but a concession of more than twice the prominence level is seen as large. A first insight into the phenomenon of sudden acceptance was given in Section 7.5.

Let us first look at plays of a weak player with a strong player. The following table provides the aggregated information on these plays.

Table 9.4: Characteristics of the weak players' shapes in plays of a weak player with a strong player

Shape	α	#	Avg steps	Avg Lo Outcome	Sudden Acceptance Lo		Final Pro-posal by Hi
					#	Avg conces-sion	
L1	55	10	4.40	34.70	10% (1)	30.00	50%
	60	3	5.00	25.00	33% (1)	20.00	67%
	70	10	4.60	21.60	30% (3)	36.33	50%
L2	55	33	12.06	33.24	9% (3)	18.33	55%
	60	23	12.52	26.96	17% (4)	25.00	57%
	70	13	10.08	24.69	23% (3)	28.67	54%
L3	55	17	4.84	35.63	100%	27.22	100%
	60	20	6.30	30.70	90% (18)	45.61	90% (18)
	70	3	7.67	19.67	100%	57.67	100%

A comparison of the agreement outcomes shows that for each α the average outcomes of the different shapes are relatively equal. The important difference between the payoffs of the different shapes is caused by the extreme difference in the bargaining length. While in shape L1 and L3 the agreement occurs in about 5 to 7 bargaining steps, the agreement in shape L2 is reached only after 12 steps. The consequence is a high discount rate which makes, ex post, the shapes L1 and L3 more favorable than shape L2.

For $\alpha = 55$ and 60 shape L1 behavior is rare, it occurs in only 16% and 7%, respectively. The high rate of revelation in the first step (38%) for $\alpha = 70$ is in accord with previous observations (see Section 8.4).

The behavior subsumed under shape L2 is the most often observed behavior (approximately in 50% of all plays). As table 9.5 shows, the revelation occurred on average in step 5 or 6. From then on the bargaining proceeds on average for additional 6 steps.

Table 9.5: Average step of revelation in shape L2

α	Gametype	Avg step of revelation in L2
55	(0, 55)	4.73
	(55, 0)	5.45
60	(0,60)	7.18
	(60, 0)	6.50
70	(0,70)	5.33
	(70,0)	7.14

The most striking feature of the L3 behavior is the high frequency of sudden acceptances by the weak player. But, at a closer look, this is the only possibility a strong player and a weak player of shape L3 can reach an agreement with an individually rational outcome for the strong player. To reach an individually rational agreement for the strong player, the weak player has to concede from a value above α to a value below $100-\alpha$. This means he has to make a sudden agreement with a concession of at least 10, 20, and 40 for $\alpha=55$, 60, and 70, respectively. Moreover, it is obvious that (in case of an individually rational outcome for the strong player) the final demand has to be made by the strong player. Remember, that for $\alpha=60$ we observed two agreements with non individually rational outcomes for the strong player. These are the two cases in which the final proposal was made by the weak player and the agreement was reached by sudden acceptance of the strong player (see table 9.4). Consider, for example, the case of $\alpha=70$. A typical situation of shape L3 is that both players demand 80 and the weak player suddenly accepts, such that his outcome is 20 with a final concession of 60.

For the other shapes it also happens that the agreement is reached by sudden acceptance by the weak player although he already revealed his type. But, this occurs rather seldom.

After looking at the features of the three shapes of behavior of the weak player, we shall study the behavior of the strong opponent. The following table gives an overview.

Table 9.6: Characteristics of the strong player in plays of the weak player and the strong player. "*Hi improves*" means that the strong player's agreement outcome is higher than his initial demand. The "*demand range*" is the difference between the maximal and the minimal demand of the player.

Shape	α	Demand range of Hi		Demand range of Lo	Hi improves
		Avg	= 0		
L1	55	5.00	50%	5.50	0%
	60	5.00	67%	5.67	0%
	70	4.50	40%	17.40	0%
L2	55	10.03	9%	28.42	15%
	60	14.30	0%	42.13	22%
	70	11.77	8%	51.69	8%
L3	55	6.84	53%	5.05	0%
	60	8.50	30%	7.50	15%
	70	9.67	33%	15.67	0%

The most interesting difference that could be found in the behavior of the strong opponent concerns the *demand range*, defined as the difference between the maximal and the minimal demand of the player. If the opponent revealed with the initial demand the strong player is "stubborn" with an average demand range of five and in half of all plays the strong player did not even make any concession (demand range=0). In case of a shape L2 behavior the bargaining mostly starts on a high demand level and then both players start to "bargain". The demand ranges of both players are very high and by subsequent concessions the weak player reveals his type. The low frequency of plays where the strong player makes no concession shows that the concession process is driven by both players. In the case that the opponent does not reveal during the negotiation process (shape L3) the demand ranges of the two players are quite similar. For $\alpha=55$, 60, and 70 the two-sided Wilcoxon matched-pairs signed-ranks test cannot reject the hypothesis that the demand range of the strong player is equal to the demand range of the shape L3 opponent at a significance level of .2.

It is remarkable that in the case of a revelation during the negotiation process the high type player is able to improve his outcome, which means that he manages to receive an agreement outcome higher than his initial demand. By the revelation of the weak opponent, which was not a priori obvious, as in L1, the strong player gains strength and is able to "exploit" the weak opponent.

We shall now look at the bargaining of two weak players for plays of games with $\alpha>50$ and remain at the distinction of the three shapes. The following three tables (9.7 to 9.9) give the number of occurrence of the three shape combinations for the different values of α. The column shape corresponds to the shape of the first mover and the row shape to the shape of the second mover.

Table 9.7: Occurrence of shape combinations for $\alpha=55$

$\alpha = 55$	L1	L2	L3
L1	3	0	3
L2	1	1	11
L3	9	11	19

Table 9.8: Occurrence of shape combinations for $\alpha=60$

$\alpha = 60$	L1	L2	L3
L1	0	0	2
L2	0	1	16
L3	4	17	23

Table 9.9: Occurrence of shape combinations for $\alpha = 70$

$\alpha = 70$	L1	L2	L3
L1	1	0	2
L2	1	2	14
L3	12	12	14

In all three tables the highest occurrence can be found on the last row and the last column. This means that the opponent did not reveal if the player revealed before. Most frequently a player of shape L2 and a player of shape L3 or two players of shape L3 play together.

The average outcome of the first mover (the row shape) is given in the following three tables (9.10 to 9.12).

Table 9.10:   Average outcome of the first mover (row shape) for $\alpha = 55$

$\alpha = 55$	L1	L2	L3
L1	55.33	–	35.67
L2	50.00	50.00	44.64
L3	53.56	58.36	44.89

Table 9.11:   Average outcome of the first mover (row shape) for $\alpha = 60$

$\alpha = 60$	L1	L2	L3
L1	–	–	32.50
L2	–	50.00	37.69
L3	64.50	60.06	50.26

Table 9.12:   Average outcome of the first mover (row shape) for $\alpha=70$

$\alpha = 70$	L1	L2	L3
L1	55.00	–	27.50
L2	51.00	55.00	45.43
L3	61.42	55.75	46.14

For all three values of α a player of shape L3 exploits the opponents of shape L1 and L2 by achieving an outcome which is considerably higher than 50.

The average number of steps that the different shape combinations need to reach an agreement is shown in the following three tables (9.13 to 9.15).

Table 9.13:   Average number of steps for $\alpha=55$

$\alpha = 55$	L1	L2	L3
L1	2.67	–	3.00
L2	3.00	7.00	6.36
L3	2.78	6.64	4.00

Table 9.14:   Average number of steps for $\alpha=60$

$\alpha = 60$	L1	L2	L3
L1	–	–	2.50
L2	–	22.00	13.63
L3	5.25	14.29	6.22

Table 9.15:   Average number of steps for $\alpha = 70$

$\alpha = 70$	L1	L2	L3
L1	3.00	–	5.00
L2	4.00	6.50	10.93
L3	4.00	12.58	3.71

As observed in the plays of a strong and a weak player, the negotiations in which a weak player of shape L2 is involved, need the largest number of steps to reach an agreement.

The phenomenon of sudden acceptance is almost only found if both weak players behave like shape L3. In this case all plays (must) end in sudden acceptance.

We shall study the characteristics of the plays of two shape L3 players in the following table.

Table 9.16:   Characteristics of the weak players in plays of two shape L3 players

Player	α	Avg outcome	Avg concession of sudden acceptor	Avg range	Range=0	Initial demand
Sudden acceptor	55	35.53	28.42	4.68	42%	68.47
	60	29.04	43.78	6.30	48%	77.09
	70	26.00	50.21	4.14	57%	80.36
Other player	55	64.47		6.37	37%	69.89
	60	70.96		8.70	30%	76.39
	70	74.00		5.57	36%	79.29

From table 9.16 one might get the impression that the bargaining behavior of the sudden acceptor is tougher than the bargaining behavior of the other player. This impression is motivated by the lower demand range of the sudden acceptor. However, for $\alpha = 55$, 60, and 70 the one-sided Wilcoxon matched-pairs signed-ranks test, applied to the average demand ranges over the six independent subjects groups, cannot reject the null hypothesis of equal demand ranges in favor of the alternative that the sudden acceptor has a lower demand range at a significance level of .1. Furthermore, for $\alpha=55$, 60, and 70 the one-sided Wilcoxon matched-pairs signed-ranks test cannot find a difference in the percentage of zero demand ranges between the sudden acceptor and the other player.

It should be noticed that for $\alpha=55$ and $\alpha=70$ nearly ⅓ (32% and 36%, respectively) of all plays of two L3 shapes ended in two steps. Typically in these plays the first mover demands for example 80, the second mover demands up to five more or less than 80, and the first mover then accepts. There is no indication why the first mover should conclude from the demand of the second mover (which can be higher as well as lower as the first mover's demand) that he is strong. For $\alpha=60$ about 17% of all plays between two shape L3 players ended in two steps.

In summary, we found that approximately half of all plays are "classical bargaining games" which start with high initial demands and reach the agreement by a concession process. With a proportion of around one third a very interesting phenomenon, which was called *sudden acceptance* was found. From a high demand level of both players, a player suddenly accepts the proposal of the opponent.

CHAPTER 10.   INDIVIDUAL ADAPTATION TO EXPERIENCE

This chapter investigates the individual reaction of a player to experience. The first part gives a qualitative description of this adaptation and the second part discusses some quantitative aspects.

10.1   QUALITATIVE DESCRIPTION OF THE ADAPTATION

Each experimental subject played 16 plays of games involving the same high alternative value α, where each gametype was experienced four times. The following analysis aims to study the effects of the past experience to the actual play. Especially, the consequences for the initial demand and the concession behavior of the players are emphasized. The characteristics describing the concession behavior are the *total concession* (= initial demand − outcome), the *average concession* (= total concession / number of demands), the *concession rate* (= 1 − (outcome / initial demand)), and finally the *concession speed* (= concession rate / number of demands).

At first we shall study the characteristics of the actual play if the previous play ended in agreement. The following table displays the relationship between the characteristic of the actual play (play i) and the previous play (play i−1).

Table 10.1:   Changes of characteristics after an agreement. Significance by the Wilcoxon matched-pairs signed-ranks test (level of .025, one-sided) is marked by "•"

Characteristic changes after an *agreement* in play i−1	Significant for α				
	30	45	55	60	70
initial demand (i) ≥ initial demand (i−1)	•	•	•	•	•
total concession (i) ≤ total concession (i−1)	•	•	•	•	•
average concession (i) ≤ average concession (i−1)	•	•	•	•	•
concession rate (i) ≤ concession rate (i−1)	•	•	•	•	•
concession speed (i) ≤ concession speed (i−1)		•		•	•
concession speed (i) ≥ concession speed (i−1)	•				

For each parameter value of α a one-sided Wilcoxon matched-pairs signed-ranks test with the six independent subject groups is conducted to examine whether the number of cases in which the relationship is true is significantly greater than the number of cases in which it is false. The cases in which the hypothesis that there is no difference in these numbers can be rejected in favor of the alternative that the relationship is true (at a significance level of .025, one-sided) are marked with a "●" in table 10.1.

If the previous play ended in agreement the initial demand of the actual play is at least as high as the initial demand of the previous play. The total concession as well as the average concession and the concession rate of the actual play are less or equal than in the previous play. The direction of the change in the concession speed differs among the different parameter values of α. For $\alpha=45$, 60, and 70 the concession speed decreases, while it increases for $\alpha=30$. For $\alpha=55$ no significant result can be found.

With the exception of the concession speed in the case $\alpha=30$ the direction of the change in the characteristics is towards a tougher bargaining behavior. This can be interpreted as the attempt to examine whether tougher bargaining leads to a "better result". The "success" of an agreement strengthens the bargaining characteristics of the players. For $\alpha=30$ the increase in the concession speed can be explained as the increasing speed to the equal split agreement.

In case of a break off in the previous play one would expect the player to be more cautious, since he experienced a failure.

Table 10.2:   Changes of characteristics after a break off. Significance by the Wilcoxon matched-pairs signed-ranks test (level of .025, one-sided) is marked by "●"

Characteristic changes after a *break off* in play i−1	Significant for α				
	30	45	55	60	70
initial demand (i) ≤ initial demand (i−1)	●	●	●	●	●
total concession (i) ≥ total concession (i−1)	●	●	●	●	●
average concession (i) ≥ average concession (i−1)	●	●	●	●	●
concession rate (i) ≥ concession rate (i−1)	●	●	●	●	●
concession speed (i) ≥ concession speed (i−1)			●	●	●

Actually, the initial demand decreases after a break off. A look at the concession process is more difficult. Since the previous play ended in break off the outcome of the player is his alternative value and cannot be taken as endpoint of the concession process. Therefore, the last demand of each player is taken instead of the outcome to evaluate the concession characteristics. Typically, the final demands of the players are not close before a break off and the total concession as well as the concession rate is not as high as in case of agreement. Thus, a lower concession rate than in the actual play might be caused by the different concession rates between plays ending in conflict and those ending in agreement.

However, in case of a break off in the previous play all relationships are in the opposite direction in comparison to an acceptance in the previous play. The bargaining behavior becomes more cautious after a break off.

Are there other characteristics of the previous play that have a different influence on the behavior of the player? How does a player react to a very long play, or how does he react to a revelation in the previous play?

A revelation of the player in the previous play as well as a revelation of the opponent does not change the relationships between the characteristics, if the previous play ended with an agreement. In case that the player revealed in the previous play and it ended in conflict, the initial demand rises.

Table 10.3:   Changes of characteristics after a break off. Significance by the Wilcoxon matched-pairs signed-ranks test (level of .025, one-sided) is marked by "•"

Characteristic change after a *break off and a revelation of the player* in play i−1	Significant for α		
	55	60	70
initial demand (i) ≥ initial demand (i−1)	•	•	•

If the player experiences the failure of a break off after a revelation by himself, he increases his demand. This is the opposite effect in comparison to a break off in general. The player might conclude that his behavior was too weak and he was too much exploitable. The other characteristics are not influenced by a break off and a revelation by the player in the previous play. Moreover, no effect is observable if the opponent revealed and the play ended in break off. Also, it does not have an influence on the relationships whether the player or the opponent initiated the break off in the previous play.

A player might also react to the length of the previous play. If it was considered as "too long", the player might decide to change his behavior in order to avoid long bargainings which are costly in terms of discounting. The quantification of "too long" is: longer than the median of all bargaining lengths the player had experienced before.

<u>Table 10.4:</u>   Changes of characteristics after an agreement. Significance by the Wilcoxon matched-pairs signed-ranks test (level of .025, one-sided) is marked by "•". *"too long"* means longer than the median of all bargaining lengths the player had experienced before

Characteristic change after a *"too long"* play $i-1$ with an *agreement*	Significant for α				
	30	45	55	60	70
concession speed (i) ≤ concession speed (i−1)	•	•	•	•	•

The player changed his concession speed after a too long play. A reason for this behavior might be that the concession speed of the player was too high in comparison to his opponent, such that the bargaining got stuck because the player was not willing to concede anymore. This may cause a long bargaining and a consequence is to lower the concession speed and to adapt it to the opponent's concession speed.

One might also suppose that the difference of the initial demands in the previous play may influence the behavior of the player in the actual play. If this difference is large, the players have different margins for concessions. This may hinder the flow of the bargaining and prolong it non-necessarily. But, no significant change in the relationships can be found.

10.2   QUANTITATIVE ASPECTS OF THE ADAPTATION

After an acceptance in the previous play a player tends to increase his initial demand and after a break off he tends to lower it. In the following this purely qualitative description of the adaptation will be complemented by the discussion of some quantitative aspects. Let $d(i)$ denote the initial demand in play i. The following table shows the average increase of the initial demand in case of agreement and the average decrease of the initial demand in case of break off for every parameter value of α. These values are calculated conditional on an increase (decrease). This means that in case of an agreement in play $i-1$ the average of $(d(i)-d(i-1))$ is calculated for all cases in which $d(i) \geq d(i-1)$, and the corresponding is true for break offs. The percentages of occurrence of these increases (decreases) are also shown in table 10.5.

Table 10.5:   Quantitative aspects of the adaptation

α	Agreement in play $i-1$		Break off in play $i-1$	
	$d(i) \geq d(i-1)$	$avg(d(i)-d(i-1))$	$d(i) \leq d(i-1)$	$avg(d(i)-d(i-1))$
30	66.45%	7.11	76.92%	-12.55
45	65.72%	9.00	69.08%	-14.00
55	74.80%	11.95	68.18%	-10.80
60	77.31%	10.18	64.51%	-9.44
70	76.09%	14.91	61.80%	-11.77

For $\alpha < 50$ the average decrease of the initial demand after a break off is considerably higher than the average increase after an agreement. A break off is a clear signal of a failure and the player performs a more drastic change. In case of an agreement the player does not know whether a tougher bargaining would have led to a higher payoff. A moderate increase of the initial demand is an attempt to examine this. For $\alpha > 50$ the average increase in case of an agreement and the average decrease in case of a break off are only slightly different. Here, a break off does not necessarily mean a failure of the bargaining behavior. Possibly both players are strong and therefore not able to agree. Hence, the decrease is more moderate and the percentage of occurrence shows that it becomes less frequent for greater α.

CHAPTER 11.   A PICTURE OF THE SUBJECTS' BEHAVIOR

In this chapter we summarize the most important results of the evaluation of the data of the game playing experiment (in the first section), and give a qualitative overall picture of the observed behavior (in the second section).

11.1   STYLIZED FACTS

In what follows we shall give an overview over the insights which could be gained on the behavior of the subjects in the game playing experiment. The important results will be pointed out in a stylized form. Detailed information about these results has been given in the previous chapters.

The prominence level:
- The prominence level is 5. More than 70% of all proposals are on prominent numbers.

The initial demands:
- The weak player's initial demand is lower than the strong player's initial demand (for $\alpha=30$, 55, 60, and 70, tested by the Wilcoxon matched-pairs signed-ranks test).
- The second mover's initial demand is lower than the first mover's initial demand (for $\alpha=45$, 55, 60, and 70, tested by the Wilcoxon matched-pairs signed-ranks test).
- The initial demands follow a decreasing trend for $\alpha<50$.
- For $\alpha<50$ there is an increasing trend in immediate equal split agreements.
- The initial demands follow an increasing trend for $\alpha>50$.
- For $\alpha>50$ the low type's initial demand is between 60 and 80 for $\alpha=55$, and between 70 and 90 for $\alpha=60$ and $\alpha=70$. The high type's initial demand is between 70 and 80 for $\alpha=55$, on 80 for $\alpha=60$, and between 80 and 90 for $\alpha=70$.

The agreement outcomes:
- There is no significant difference in the agreement outcome of the first mover and the second mover (Wilcoxon matched-pairs signed-ranks test).
- The majority of the agreement outcomes is on prominent numbers.

- For $\alpha < 50$ the agreement outcomes converge to the equal split (order test). The average deviation from the equal split is below 1 in the fourth experience level.

- For $\alpha = 55$ the agreement outcomes become more asymmetric with a higher level of experience (order test). This is especially true for plays of two weak players.

- For $\alpha = 70$ there is a considerable number of equal split agreements in plays of two weak players.

- The most successful of the tested prediction concepts for the agreement outcomes is the equal split for $\alpha < 50$, and for $\alpha > 50$ the set of the prominent numbers in the following area: the strong player receives at least $\alpha + 5$, and the weak player's lower outcome bound is the equal split of the difference, adjusted to the adjacent prominent number below this value.

The average payoff:

- On average, the high type players are not able to achieve individually rational payoffs for $\alpha > 50$. This means that the average payoff of the high type first mover is below α and the average payoff of the high type second mover is below $\delta\alpha$. Especially, this contradicts the assumption that Nash equilibria are played.

- An increase in the number of immediate break offs is not a consequence of this experience.

- For $\alpha < 50$ the high type players receive, on average, individually rational payoffs.

- Clearly, the low type players receive individually rational payoffs.

The revelation: $(\alpha > 50)$

- In plays of the weak and the strong player a revealing weak player receives on average a higher payoff than a non-revealing weak player.

- In plays of two weak players a revealing weak player receives on average a lower payoff than a non-revealing weak player.

- Overall, a non-revealing weak player has a higher average payoff than a revealing weak player, for $\alpha = 55$ and $\alpha = 60$. For $\alpha = 70$ the revealer has the higher payoff.

- If both players reveal the first revealer's outcome is equal or higher than the second revealer's outcome.

- The strong player has a higher payoff playing a revealing player than playing a non-revealing player.

- The agreement rate in a play of a strong player and a weak player is strikingly higher if the weak player has revealed than in case he has not revealed.
- The ratio of revealer to non-revealer is about 1 for $\alpha=55$, and about ½ for $\alpha=60$ and $\alpha=70$.
- The revelation occurs roughly in step 4, and the play ends after 5 additional steps.

The break offs:
- The conflict frequency increases with the shrinkage of the surplus.
- There is no significant trend in the conflict frequencies of the different experience levels for $\alpha=30$, 55, 60, and 70 (order test).
- There is a decreasing trend in the conflict frequencies of the different experience levels for $\alpha=45$ (order test).
- Immediate break offs are not of importance for $\alpha>50$.
- If the weak player initiated the break off, in more than 80% of the cases the opponent did not offer at least 50 during the bargaining process.
- If the strong player initiated the break off, in the majority of all cases the opponent did not offer an individually rational amount to the player during the bargaining process.
- For $\alpha<50$ other reasons for break offs by the strong player are that the opponent did not offer at least 50 ($\alpha=30$), and that the opponent did not offer more than 50 ($\alpha=45$).
- For $\alpha>50$ another, but less important, reason for a break off of the strong player is that the opponent did not offer at least $\alpha+5$.
- If the weak opponent did not offer an individually rational amount to the strong player and did not reveal his type, the strong player breaks off after 5 steps, for $\alpha>50$.
- If the weak opponent did not offer an individually rational amount to the strong player but revealed his type, the strong player breaks off after 8 to 15 steps, for $\alpha>50$. After the revelation the weak player mostly proposes the equal split.

The bargaining length:
- The bargaining between a weak and a strong player proceeds on average longer than the bargaining between two equally strong players.
- The bargaining length of the plays ending in conflict is approximately twice as long as the length of plays ending in agreement.
- For $\alpha=45$ the bargaining length decreases with experience (order test).

- For $\alpha=60$ the bargaining length increases with experience (order test).

The concession behavior:

- A player almost always makes a positive concession.
- The concession process of a player is decreasing.
- The numbers divisible by 10 are "points of stronger resistance". If the previous demand exceeded such a number by $1 \leq r \leq 4$, the actual concession is at most r.
- For $\alpha=30$, 45, and 55, the numbers divisible by 10 are "points of stronger resistance". If the previous demand exceeded such a number by $1 \leq r \leq 3$, the actual concession is at most r.
- The number 50 is a "point of stronger resistance". If the previous demand exceeded such a number by $1 \leq r \leq 2$, the actual concession is at most r.
- For $\alpha=60$ and 70, the equal split of the surplus in addition to α, rounded to the next higher prominent number is a "point of stronger resistance". If the previous demand exceeded such a number by $1 \leq r \leq 4$, the actual concession is at most r.
- The points of stronger resistance are interpreted as natural choices of planned aspiration levels of the subjects.

The bargaining process:

- For $\alpha < 50$ there is a positive correlation between the initial demand and the concession rate and the initial demand and the number of large concessions.
- For $\alpha < 50$ the bargaining process follows an exponential process.
- For $\alpha > 50$ about half of all plays are "classical bargaining games" where the players start with high initial demands and approach an agreement by successive concessions of both players. In about one third of the plays the phenomenon of sudden acceptance was observed. From a high demand level of both players, one player suddenly accepts the proposal of the opponent.

The individual adaptation to experience:

- If the previous play ended in agreement the
  - initial demand increases,
  - total concession decreases,
  - average concession decreases,
  - concession rate decreases.

- If the previous play ended in conflict the
  - initial demand decreases,
  - total concession increases,
  - average concession increases,
  - concession rate increases.

- If the previous play ended in agreement and was "too long" the concession speed decreases.

- If the previous play ended in conflict and the player revealed in the previous play the initial demand increases ($\alpha = 55$, 60, 70).

- For $\alpha < 50$ the average decrease of the initial demand after a break off is considerably higher than the increase after an agreement. Here, a break off is a clear signal of a failure.

- For $\alpha > 50$ the average decrease after a break off is not significantly different from the average increase after an agreement in the previous play. This is because a break off is not necessarily a signal of a failure. Break offs are non-avoidable if both players are strong.

11.2   A QUALITATIVE PICTURE OF SUBJECTS' BEHAVIOR

In the previous section the main results of the evaluation of the experimental data were summarized in a stylized form. Now we shall give a qualitative picture of the behavior in the game playing experiment and its interpretation.

GAMES WITH $\alpha < 50$

Experienced players agree on the equal split, independently of their alternative values. The behavior of naive players is less uniform. They make higher initial demands in order to reach more asymmetric outcomes. But, they also have a higher conflict rate and a longer bargaining length (which means that the discount factor is higher). With experience the players learn that a high type does not necessarily achieve a higher payoff than a low type since he has no instrument to prove that he is actually strong. This leads the players to the prominent solution of the equal split, the egalitarian solution concept. They reach this solution immediately, or after a few steps. Deviations from the equal split solution are very small (on average below 1) and the agreement rate increases. The facts that the incomplete information makes the two types "indistinguishable" and that the equal split solution exceeds the alternative values of both types of players favor this focal point as a solution concept.

GAMES WITH $\alpha > 50$

For $\alpha < 50$ the players learned by experience that the weak player and the strong player have the same bargaining power. A similar result is true for the plays of games with $\alpha > 50$. Of course, the strong player can always guarantee himself a high payoff (his discounted alternative value) by breaking off. But, the incomplete information weakens the bargaining power of the strong player. He is not able to prove his strong position during the bargaining game and therefore may always be seen as a weak player who imitates a strong player. The consequence for his agreement outcome is that it is significantly lower than in a comparable game with complete information. However, the weak player learns that he is able to reach highly asymmetric agreements in a play of two weak players by imitating a strong player. The initial demands increase with experience, and the agreement outcomes in a play of two weak players become more asymmetric with experience. The weak player is able to achieve outcomes nearly in the whole interval, but interestingly they are not distributed equally. They show major peaks on the prominent numbers, which are the numbers divisible by 5 without remainder. The tendency to choose prominent numbers is also found for strong players. The most successful of the tested prediction concepts for the agreement outcomes is the set of the prominent numbers in the following area: the strong player receives at least $\alpha+5$, and the weak player's lower outcome bound is the equal split of the difference, adjusted to the adjacent prominent number below this value. On average, the high type players are not able to achieve individually rational payoffs. This means that the average payoff of the high type first mover is below α and the average payoff of the high type second mover is below $\delta\alpha$. Especially, this contradicts the assumption that Nash equilibria are played. An increase in the number of immediate break offs as a consequence cannot be found.

Weak players do not always follow the strategy of an imitation of the strong player. One half to one third of the weak players reveal their type by proposing a value less than α. In a play with a strong player this is a very successful tactic. The agreement rate is extremely high and both players receive a higher average payoff compared to a play of a strong player and a weak player who did not reveal. But, in case the opponent is weak too, the revealer is exploited. He receives a very low payoff such that over all plays a revelation is not advantageous in terms of expected payoff (for $\alpha=55$ and 60). For $\alpha=70$, however, a revelation is advantageous. Why this? For very high alternative values of α the potential additional gain of the high type is small, therefore he might be more likely to be willing to end the game with break off, if he sees no chance for an agreement. A revelation of the weak player signals that there is a positive surplus to allocate, and possibly the players can agree on an allocation. A revelation in case of a weak opponent often leads to the equal split agreement. A considerable number of weak opponents does not exploit the revelation of the weak player and receive a high outcome, but accept the proposal of the equal split. The higher likelihood of a break off by the strong player makes the weak players more cautious.

The bargaining process is a succession of (positive) concessions by the players. The process starts with large concessions, but they shrink in the course of the bargaining. There are values of special importance in the bargaining process. These values are "points of stronger resistance", which means that they are not approximated with the same concession speed as other values. These values are all numbers divisible by 10 without remainder and, among them, the value 50 plays an even more outstanding role. Players hesitate to jump over these values in the concession process, although the average concession from the previous demand would lead to a lower value. These "points of stronger resistance" can be seen as natural choices for planned aspiration levels of the players. A player plans to achieve a planned aspiration level as the outcome and therefore approaches it very carefully, but as he realizes that it is not achievable, he abandons it and forms another aspiration level. We already pointed out the strong influence of prominence in the agreement outcomes. All numbers on the prominence level 10 serve as possible planned aspiration levels of the players. The outstanding importance of the value 50 is explained by its focal role as the equal split. Also in games with $\alpha > 50$ a weak player hopes to obtain this value, since it is the egalitarian allocation of two weak players. Since a weak player does not know whether his opponent is weak, too, he has to "defend" the equal split as long as it is possible. Once the demand is lowered below 50, there is no canonical distribution for two weak players anymore. This gives the opponent, even if he is weak, the possibility to obtain a considerably higher outcome than 50.

Only about half of all plays are "classical bargaining games" where the players start with high initial demands and approach an agreement by successive concessions. In about one third of the plays the phenomenon of sudden acceptance was observed. From a high demand level of both players, one player suddenly accepts the proposal of the opponent. If sudden acceptance is found in a play of a strong and a weak player, it is the weak player who suddenly accepts. He pretended to be the strong player, but now strongly beliefs that his opponent is strong. In order to avoid a break off he suddenly accepts the proposal of the opponent.

In games with $\alpha > 50$ two strong players cannot agree on individually rational agreements. Therefore, the strong player has to break off the bargaining as soon as he strongly believes that his opponent is strong, too. If a strong player decides to break off the bargaining, in the vast majority of the cases the opponent did not offer an individually rational amount. The player concludes that the bargaining is not worthwhile anymore since the opponent is either strong too, or a stubborn weak player. But, when is the right time to break off? How does the player know that the opponent does not accept in the next step? A break off of the strong player in case the opponent did not offer an individually rational amount occurs approximately in step 5. In step 4, roughly, a revelation of the weak player occurs. Taking these average values, a revealing weak player would have revealed before the break off of the strong player. If the opponent revealed, but nevertheless did not offer an individually rational amount, the break off of the strong player occurs after 8 to 15 steps. In this case the opponent mostly proposes the equal split, strongly believing that his opponent is weak, too.

The reason for a weak player to initiate a break off is that the opponent did not offer at least 50 during the bargaining process.

Plays ending with break off take about twice as many steps as plays ending in agreement.

If the player experienced a break off in the previous play he lowers his initial demand and increases the concessions. In case the player was weak in the previous play, a break off is (mostly) a clear signal of a failure. Either the initial demand was too high or the concessions too low to come to an agreement. Therefore the player decides to lower the initial demand and to increase the concessions. In case that the previous play ended in agreement the player reacts exactly in the opposite way. He increases his initial demand and adopts a tougher concession behavior. The player experienced that with the behavior of the previous play an agreement was possible. Now he wants to figure out whether an agreement is also possible under the condition of a tougher bargaining. Since for $\alpha < 50$ the signal of the failure by a break off is stronger, the adaptation after a break off is much stronger than the adaptation after an agreement. The player only carefully examines whether a tougher behavior is successful, too. For $\alpha > 50$ a break off is no longer a clear signal of failure. The increase of the initial demand after an agreement is approximately as large as the decrease after a break off in the previous play.

CHAPTER 12.   THE STRATEGY EXPERIMENT

12.1   ORGANIZATION OF THE STRATEGY EXPERIMENT

In the Winter term 1991/92 we conducted a strategy experiment for the two-person bargaining problem with incomplete information at the Bonn Laboratory of Experimental Economics in the framework of a student's seminar. A strategy experiment is designed to gain strategies for a certain problem from highly experienced subjects. Game playing experiments, in contrast, explore the spontaneous behavior, and even in the case of experienced subjects the experimenter only observes the actions of the subjects and is not able to identify the strategies they emerge from.

A strategy experiment starts with an experience phase where the subjects experience the problem in subsequent game playing sessions. The experience phase is followed by the development of a strategy for the problem, and its improvement after several tournaments. The strategy experiment lasted for about four month.

The topic of the strategy experiment was the two-person bargaining game with incomplete information, as described in Section 2.1, with a discount factor of $\delta = .99$ and a smallest money unit of $\mu = 1$. In particular, this means that, like in the game playing experiment, the feasible demands are the integer values in the interval $[0, 100]$. The high alternative values were chosen as 30, 45, and 60, and in contrast to the game playing experiment the players had to experience all three possible values for α. From a player's point of view four different situations are distinguishable for each α. The player may have the low or the high alternative value, and he may be the first or the second mover. Since in each situation, the opponent may be either weak or strong, the player can be in a total of eight different situations for each α. Therefore 24 situations were possible for each player, namely eight for each value of α, where the player was able to distinguish twelve of them (four for each α).

The participants of the strategy experiment were 32 economics students which had passed the middle exams in economics and were familiar with a programming language. They never participated in two-person bargaining games or in a strategy experiment before. At the end of the seminar they received a graded certificate, with grades largely based on the success in the strategy experiment. This certificate is a part of the admittance to the final exams. The grade of a participant was determined by his performance in the introductory game playing sessions, the success of his strategy in the tournaments, and the essay describing the development and the reasoning of the strategy. The main emphasis was on the last tournament.

The 32 participants were distributed in 4 groups of 8 each and all seminar activities except the introductory sessions and the final tournament took place within these 4 groups. An identification number was assigned to each subject and in the evaluation of the experimental data we shall refer to the identification number instead of the name of the subject. The subjects with the id numbers 1 to 8 formed the first group, the second group was formed by the subjects with the id numbers 9 to 16, and so on.

The strategy experiment started with a survey on the programming language Turbo Pascal, consisting of theoretical lectures and exercises in the computer laboratory. This obligatory part took the first four weeks of the experiment and ensured that all participants were able to handle the basic features of Turbo Pascal.

In parallel to the programming course the game playing sessions took place. In four sessions the participants experienced the 24 possible situations. In each session a subject played changing opponents of his group anonymously. Each session consisted of six plays: two plays of games with $\alpha=30$ were followed by two plays of games with $\alpha=45$ and two plays of games with $\alpha=60$. The subjects were informed in advance about this succession, and in addition in each of these plays they were informed about the value of α, such that there was no misconception about the high alternative value of the actual play. Moreover, they were informed about the 24 possible situations and that the four sessions together cover these 24 situations in a random order. The further introduction to the rules of the two-person game with incomplete information was identical to the one in the game playing experiment reported in Part II (see Appendix A). The same software was used to conduct the game playing sessions.

After each session each participant received an overview over all bargaining results of his group. It consists of the bargaining outcomes, the number of steps needed to reach an outcome and the respective discount of each of the 24 (= ½ · 8 players · 6 plays per player) plays of each session. The players were ignorant about the actual bargaining partners of each reported play. In addition, each player received an individual result sheet, where a mark indicated each play the player was involved. Moreover, the players received a ranking which showed the payoff sum of each group member and the position of the player in the ranking of these payoffs. After the fourth session these payoffs were comparable since all players played the same 24 games.

After the experience phase the participants started the programming of the strategies. In each group two tournaments, restricted to the group members, followed two programming periods of three weeks each. Three weeks after the second group tournament the final tournament of all strategies was conducted. Due to illness one participant (id number 12) had to leave the experiment after the second tournament, and therefore only 31 strategies were available for the final tournament. With the final strategies we also conducted a third tournament within the four groups in order to compare the performance of the final strategy of each participant in the subject group as well as in the whole seminar population.

For each of the three tournament dates the participants had to submit a strategy for the bargaining problem in the form of a Turbo Pascal program. In the two group tournaments each strategy of a group member played each other strategy of a group member in each of the 24 situations. This means that each strategy played 168 (= 7 other members · 24 situations) plays, such that in each group tournament 672 (= ½ · 8 members · 7 other members · 24 situations) plays were involved. After each tournament the players received the complete bargaining processes of the 168 plays, the own strategy was involved and the bargaining results of the 672 plays of the group. The players were ignorant about the developers and the contents of strategies they played. Like in the game playing sessions a ranking of the payoff sums was given to each group.

In the final tournament all 31 strategies played each other in each of the 24 situations, such that there was a total of 11160 (= ½ · 31 participants · 30 other participants · 24 situations) plays in this tournament. Here the players received an overview over the tournament for each of the 24 situations (the average outcomes, the average bargaining lengths, and the number of agreements) and the data of their strategy in each situation. A ranking of the payoffs indicated the position of each strategy in each situation. A more detailed listing of the bargaining results was not possible due to the large quantity of data.

The participants had to report the development of their strategy and the experience in the different tournaments in a seminar paper, which also had to contain a detailed description and explanation of their strategy.

In a tournament of two strategies the problem of an "infinite" play may occur, if both strategies do not specify a termination rule. To overcome this problem we instructed the subjects that a play would be terminated at an arbitrary step after the discounted coalition value is too low to be relevant for the success in the tournament. Actually, we chose step 917, where the discounted value of 100 is lower than .01, below the computational precision.

The method of conducting a strategy experiment was developed in Selten (1967b). A strategy experiment for a Cournot duopoly was conducted by Selten, Mitzkewitz, and Uhlich (1988). Keser (1992) investigated strategies for a duopoly with demand inertia. The strategy planners were not experienced in playing the game in spontaneous plays, but received the results of game playing sessions played by different subjects.

12.2   RESULTS OF THE GAME PLAYING SESSIONS

This section summarizes the results of the introductory game playing sessions in the experience phase of the strategy experiment. Each player was involved in 24 plays, exactly one for each possible situation. From the experimenter's point of view there are only four and not eight different situations for each α. The situation where the player is weak and the first mover and the opponent is strong and the second mover coincides, from the experimenter's point of view, with the situation where the opponent is weak and the first mover and the player is strong and the second mover. This reduces the number of situations, from the experimenter's point of view, to four for each α. The experimental setup was designed such that the 24 plays are two successions of the 12 situations distinguishable from the experimenter's point of view. In the first two weeks the first set of 12 situations was played and in week three and four the participants played the second set. This means we can identify two experience levels, which we shall shortly call the *inexperienced* and the *experienced* phase of the 12 situations.

For the report of the bargaining results we shall further aggregate the situations $(0,\alpha)$ and $(\alpha,0)$. This means in the case of a game of a strong player and a weak player we shall neglect the fact who moved first since this seems to be of no influence for the results. Also in the game playing experiment (see Chapter 6) we found this phenomenon. Finally, we shall aggregate the data over the four subject groups. Since they do not show significant differences this will be done for the sake of brevity.

The following three tables (12.1 to 12.3) provide the distribution of the agreement outcomes, the number of conflicts, and the average number of steps needed to reach an agreement and a conflict, respectively, in the inexperienced and the experienced phase. For the case that both players have the same alternative the table shows the higher of the two agreement outcomes. In the other case the outcome of the player with the high alternative is given. In case both players have the same alternative value we observed 16 plays in the first as well as in the second phase, in the other case 32 plays were observed in each of the two phases (16 from the case that the strong player was the first mover and 16 from the case that the weak player was the first mover).

Table 12.1:   Distribution of the agreement outcomes in the plays of games with $\alpha=30$

Number of agreement outcomes on...	Inexperienced			Experienced		
	(0,0)	(30,0)	(30,30)	(0,0)	(30,0)	(30,30)
50	13	20	7	16	28	9
51	1	1				
53			1			
55	1					
60		1	1		2	
65		3				
66		1				
70		2				
Avg steps in agreement	4.07	6.66	10.11	1.75	2.33	3.56
Number of conflicts	1	4	7	0	2	7
Avg steps in conflict	9.00	12.33	4.29	–	10.5	9.29

Table 12.2:   Distribution of the agreement outcomes in the plays of games with $\alpha=45$

Number of agreement outcomes on...	Inexperienced			Experienced		
	(0,0)	(45,0)	(45,45)	(0,0)	(45,0)	(45,45)
50	11	15	5	15	18	6
51	2	2				
53	1					
55	1			1		
60		1				
65	1					
74					1	
Avg steps in agreement	5.81	4.94	3.40	3.19	1.95	2.67
Number of conflicts	0	14	11	0	13	10
Avg steps in conflict	–	7.21	6.45	–	6.23	3.90

Table 12.3:   Distribution of the agreement outcomes in the plays of games with $\alpha=60$

Number of agreement outcomes on...	Inexperienced			Experienced		
	(0,0)	(60,0)	(60,60)	(0,0)	(60,0)	(60,60)
50	8			11	2	
60	1			1		
65	4			1		
67	1					
69					1	
70	1	2		1		
75		1		1		
76	1					
80		1		1		
90		1				
Avg steps in agreement	6.56	3.20	—	5.00	1.67	—
Number of conflicts	0	27	16	0	29	16
Avg steps in conflict	—	6.22	3.38	—	4.07	2.25

For $\alpha<50$ the tables show a clear preference for the equal split. Even if there are few agreements different from the equal split in the first experience phase, the plays of the experienced players give a clear signal. If an agreement is reached then almost always it is the equal split agreement. The number of conflicts does not shrink significantly with the experience and is considerably high (about 40%) for the plays of games with $\alpha=45$. However, with experience the players shorten the number of steps they need to agree. The number of steps of plays ending in conflict is unchanged and very high.

For $\alpha>50$ the vast majority of plays of two experienced low type players ends with the equal split. This result is not as clear as in the plays of games with $\alpha<50$, which is also demonstrated by the relatively high number of 5 steps needed to reach the agreement in the high experience phase. A low type and a high type very rarely agree, but this after a small number of steps. In contrast to the game playing experiment we do not observe the highly asymmetric outcomes in plays of two weak players. There the asymmetry even rises with experience, while the participants of the strategy experiment become more egalitarian with experience. The subjects of the game playing experiment played just one session with completely anonymous opponents. They were motivated by monetary gains. The participants of the strategy experiment played for four sessions with anonymous negotiation partners of a fixed group. The incentive was to gain experience with the bargaining problem and to achieve a high number of points as a part of the final grade. This difference in the relationship of the subject group and the incentive structure may lead to these more egalitarian bargaining results. However, an alternative explanation is that the participants of the strategy experiment transferred the experience with equal split agreements in the case of $\alpha < 50$ to the case of $\alpha > 50$. Remember, that they experienced all parameter values of α, while the subjects of the game playing experiment experienced only one parameter value of α.

In the game playing experiment we saw that for $\alpha > 50$ the high type players were on average not able to receive an individually rational payoff. The average payoff of the high type first mover was below α and the average payoff of the second mover mostly was below $\delta\alpha$. In the following we shall verify whether the same is true for the strategy experiment.

Table 12.4:   Average payoffs of the different types of players

α	Type	Inexperienced		Experienced	
		First mover	Second mover	First mover	Second mover
30	L	40.23	43.19	47.64	47.63
	H	41.14	45.51	43.74	43.78
45	L	39.35	35.00	38.55	39.05
	H	44.76 •	45.72	47.25	46.02
60	L	25.93	25.01	23.16	28.97
	H	59.57 •	58.88 •	58.66 •	58.60 •

The table gives the average payoff of the different types in the different mover positions for the three parameter values of α. A "•" indicates an average payoff which is not individually rational. The observation goes in the same direction as in the game playing experiment. For $\alpha = 60$ both high types are, on average, not able to achieve the payoff they could guarantee themselves by an immediate break off.

12.3   RESULTS OF THE GROUP TOURNAMENTS

In this section we shall report on the results of the three group tournaments. Tournament 1 and tournament 2 are the intermediate tournaments during the strategy experiment and tournament 3 is the group tournament with the final strategies of the participants. The development in each group will be reported by the average payoffs of the four different types of players for every α.

Table 12.5:   Average payoffs of the types in group tournament 1

α	P_L^1	P_L^2	P_H^1	P_H^2
	Group 1			
30	49.34	49.33	48.26	48.26
45	43.14	40.47	46.73	47.11
60	24.32	28.30	59.15 •	60.70
	Group 2			
30	42.15	41.59	46.08	46.21
45	35.29	36.30	46.99	46.90
60	22.63	22.43	59.95 •	59.79
	Group 3			
30	49.94	49.97	50.00	49.97
45	49.90	49.91	49.96	49.95
60	30.34	28.77	60.38	60.14
	Group 4			
30	49.53	49.97	49.72	49.53
45	42.11	43.16	46.80	46.25
60	28.85	26.92	57.80 •	60.26

Remember, that P_L^1 denotes the average payoff of a weak first mover, P_L^2 denotes the average payoff of a weak second mover. P_H^1 and P_H^2 denote the average payoff of a strong first and second mover, respectively. A "•" marks an average payoff which is not individually rational.

Except for the third group the high type first movers were, on average, not able to achieve individually rational payoffs for $\alpha=60$.

Table 12.6:   Average payoffs of the types in group tournament 2

α	P_L^1	P_L^2	P_H^1	P_H^2
	Group 1			
30	50.00	50.00	50.00	50.00
45	49.18	49.55	49.61	49.48
60	27.17	27.37	58.24 •	60.23
	Group 2			
30	41.29	41.19	45.97	46.25
45	38.04	38.42	47.59	47.66
60	19.35	18.88	58.34 •	57.09 •
	Group 3			
30	50.00	50.00	50.00	50.00
45	50.00	49.97	49.94	49.97
60	32.87	29.83	59.84 •	59.87
	Group 4			
30	50.00	50.00	50.00	50.00
45	44.91	45.85	48.90	47.84
60	30.99	30.22	60.50	60.88

Like in the first tournament, in three of the four groups the high type first movers were, on average, not able to achieve individually rational payoffs for $\alpha=60$. Moreover, this is true for the high type second mover of group 2. In groups 1, 3, and 4 all plays of $\alpha=30$ ended immediately on the equal split agreement. Furthermore, two weak players agreed immediately on the equal split for $\alpha=45$, in group 3. In general, there is a slightly increasing tendency in all average payoffs, which means that the strategies played more efficiently in the second tournament.

Table 12.7:   Average payoffs of the types in group tournament 3

α	P_L^1	P_L^2	P_H^1	P_H^2
	Group 1			
30	50.00	50.00	50.00	50.00
45	49.44	49.91	49.90	49.83
60	28.14	28.44	60.06	60.39
	Group 2			
30	37.28	37.28	37.28	37.28
45	36.77	37.20	37.16 •	37.12 •
60	19.64	19.47	43.43 •	44.39 •
	Group 3			
30	50.00	50.00	50.00	50.00
45	50.00	49.97	49.94	49.97
60	30.61	29.00	60.55	60.53
	Group 4			
30	50.00	50.00	50.00	50.00
45	43.29	43.57	48.01	47.54
60	32.75	30.64	62.77	62.52

The final strategies yield only for the second group non individually rational average payoffs, and this for both high types, for $\alpha=45$ and $\alpha=60$. The groups 1, 3, and 4 remained at the immediate equal split agreements for all plays of games with $\alpha=30$. In comparison to the first tournament there is a clear tendency towards higher efficiency in the strategies of the groups 1 and 4. Group 2 is distinguished by very low average payoffs in all situations.

In the seminar paper the participants had to report the development of their strategy during the tournaments. A common statement was that the strategies developed towards a more moderate and soft bargaining. By a high number of conflicts or long negotiations the players gained the insight that a softer (than the initially very tough) bargaining would be more profitable. This leads to more moderate demands and a weaker acceptance behavior.

12.4   RESULTS OF THE FINAL TOURNAMENT

The results of the tournament which was conducted with all 31 final strategies are contained in the following four tables. For each strategy the average payoffs of the four possible types for the three parameter values of α is given. These four average payoffs are the average payoff of the weak types L_1 (P_L^1) and L_2 (P_L^2) and the strong types H_1 (P_H^1) and H_2 (P_H^2). Moreover, theses tables show the rank of the strategy according to this average payoff. Rank 1 is assigned to the strategy with the highest and rank 31 to the strategy with the lowest average payoff. Equal payoffs receive the mean rank. Since the average payoffs are displayed with three digits, it may happen that two average payoffs that look the same have different ranks. But, actually the payoff with the lower rank is the greater one.

For $\alpha=30$ all first movers receive an average payoff of exactly 50, except for strategies 11 and 14. Strategy 11 receives a payoff slightly higher than 50, and therefore rank 1 is assigned to this strategy. Strategy 14 has the worst performance as a first mover. However, strategy 14 is the most successful one as a second mover. As a second mover no player is able to achieve 50 as an average payoff. The strategies 11 and 14 are the only ones which do not demand exactly 50 as a first mover. They demand 65 and 63, respectively. As a second mover all strategies accept a proposal of at least 50. Therefore, the first mover's proposal of the equal split is accepted immediately, which explains the average payoffs of 50 for all but two first movers. The lower payoff as a second mover is caused by the two strategies which do not propose the equal split as a first mover, such that a longer bargaining is necessary to achieve an agreement. This causes a discounting of the outcomes.

For $\alpha=45$ the high type first mover of strategy 26 and both high types of strategy 27 are not able to achieve an individually rational average payoff, while the other strategies do. For $\alpha=60$ there are ten strategies which do not achieve an individually rational payoff as a high type first mover. These are all the strategies with a rank greater or equal to 22. Only strategy 5 fails to achieve an individually rational payoff for the high type second mover.

Table 12.11 shows the average payoff of the whole tournament and the rank which corresponds to this payoff. The left part of the table shows this information sorted by the strategy number, and the right part shows it sorted by the rank.

Table 12. 8:   Average payoffs of the types in the final tournament for $\alpha = 30$

Strategy	P_L^1	Rank	P_L^2	Rank	P_H^1	Rank	P_H^2	Rank
1	50.000	16	49.934	17	50.000	16	49.934	17.5
2	50.000	16	49.934	17	50.000	16	49.934	17.5
3	50.000	16	49.934	17	50.000	16	49.934	17.5
4	50.000	16	49.934	17	50.000	16	49.934	17.5
5	50.000	16	49.934	17	50.000	16	49.934	17.5
6	50.000	16	49.934	17	50.000	16	49.934	17.5
7	50.000	16	49.869	28	50.000	16	49.297	30
8	50.000	16	49.934	4.5	50.000	16	49.934	4.5
9	50.000	16	49.934	4.5	50.000	16	49.934	4.5
10	50.000	16	49.934	4.5	50.000	16	49.934	4.5
11	50.000	1	49.950	2	50.000	1	49.950	2
13	50.000	16	49.934	4.5	50.000	16	49.934	4.5
14	48.477	31	49.967	1	48.901	31	49.967	1
15	50.000	16	49.934	17	50.000	16	49.934	17.5
16	50.000	16	49.934	17	50.000	16	49.934	17.5
17	50.000	16	49.934	17	50.000	16	49.934	17.5
18	50.000	16	49.934	17	50.000	16	49.934	17.5
19	50.000	16	49.934	17	50.000	16	49.934	17.5
20	50.000	16	49.297	30	50.000	16	49.934	17.5
21	50.000	16	49.855	29	50.000	16	49.855	29
22	50.000	16	49.934	17	50.000	16	49.934	17.5
23	50.000	16	49.934	17	50.000	16	49.934	17.5
24	50.000	16	49.934	17	50.000	16	49.934	17.5
25	50.000	16	49.100	31	50.000	16	49.100	31
26	50.000	16	49.934	17	50.000	16	49.934	17.5
27	50.000	16	49.934	17	50.000	16	49.934	17.5
28	50.000	16	49.934	17	50.000	16	49.934	17.5
29	50.000	16	49.934	17	50.000	16	49.934	17.5
30	50.000	16	49.934	17	50.000	16	49.934	17.5
31	50.000	16	49.934	17	50.000	16	49.934	17.5
32	50.000	16	49.934	17	50.000	16	49.934	17.5

Table 12.9:   Average payoffs of the types in the final tournament for $\alpha=45$

Strategy	P_L^1	Rank	P_L^2	Rank	P_H^1	Rank	P_H^2	Rank
1	43.455	30	48.037	8	49.038	24	49.154	22
2	46.667	21	46.517	29	49.578	7	49.399	7
3	46.667	21	46.517	29	49.464	16.5	49.314	14.5
4	46.667	21	46.517	29	49.491	15	49.335	12
5	48.489	4	49.031	3	49.210	22	49.031	25
6	47.909	9	47.209	17	49.527	13	49.356	10
7	48.088	7	47.865	11	49.541	10	49.117	23
8	46.667	21	46.517	23.5	49.637	4.5	48.583	27
9	46.184	28	48.093	5.5	49.141	23	49.549	1
10	46.667	21	46.517	23.5	49.527	13	49.399	5
11	45.643	29	48.093	7	49.349	21	49.549	2
13	46.667	21	46.517	23.5	49.578	7	49.229	21
14	42.193	31	47.253	13	47.904	27	49.422	3
15	46.667	21	46.517	29	49.527	13	49.399	7
16	47.276	11.5	47.209	17	49.464	16.5	49.314	14.5
17	46.667	21	46.526	20	48.898	25	49.104	24
18	46.667	21	47.120	19	49.464	19	49.299	20
19	47.276	11.5	47.209	17	49.637	4.5	49.311	17
20	47.475	10	47.252	14	49.549	9	49.303	19
21	48.284	5	47.270	12	49.725	1	49.410	4
22	46.667	21	46.517	29	49.464	19	49.314	14.5
23	46.667	21	46.517	23.5	49.578	7	49.399	7
24	46.667	21	46.517	23.5	49.464	19	49.314	14.5
25	49.505	1	47.250	15	49.670	2	48.650	26
26	48.927	2	48.763	4	44.566	31	44.568	30
27	47.001	13	49.392	1	44.691	30	44.496	31
28	46.789	14	47.996	9	45.609	29	45.060	29
29	47.924	8	48.093	5.5	45.888	28	45.690	28
30	46.667	21	46.517	23.5	49.527	11	49.356	9
31	48.185	6	47.934	10	49.640	3	49.346	11
32	48.489	3	49.039	2	48.886	26	49.309	18

Table 12.10:   Average payoffs of the types in the final tournament for $\alpha=60$

Strategy	P_L^1	Rank	P_L^2	Rank	P_H^1	Rank	P_H^2	Rank
1	26.528	25	29.405	8	59.823	23	60.268	14
2	27.695	20	28.255	12	60.000	19.5	60.000	20.5
3	27.995	18	25.957	26	61.180	3	61.051	5
4	30.262	5	30.350	4	61.211	2	61.294	2
5	24.453	30	29.260	9	59.510	26.5	59.024	31
6	28.018	17	27.402	19	59.510	26.5	59.684	28
7	29.712	8	29.102	10	60.116	16	60.378	13
8	29.337	10	28.162	13	60.436	12	60.094	16
9	27.630	21	27.949	15	60.000	19.5	60.000	20.5
10	24.658	29	24.805	30	59.823	24	59.839	25
11	24.685	28	25.469	28	59.510	26.5	59.684	27
13	24.974	27	25.144	29	60.000	19.5	60.000	20.5
14	28.816	14	26.597	25	55.092	31	60.529	10
15	23.126	31	23.817	31	60.294	13	60.046	17
16	29.389	9	28.434	11	59.500	30	60.000	20.5
17	28.211	15	31.298	2	60.608	10	60.568	9
18	29.048	13	27.451	18	60.115	17	59.465	30
19	29.086	12	28.143	14	59.510	26.5	59.867	24
20	26.872	24	27.118	22	60.888	6	60.512	11
21	31.406	4	27.063	23	60.000	19.5	60.000	20.5
22	27.804	19	27.357	20	60.609	9	60.453	12
23	31.723	3	29.824	6	60.171	15	59.655	29
24	27.176	23	26.964	24	61.097	4	60.620	7
25	31.723	2	31.125	3	59.500	29	60.000	20.5
26	26.481	26	25.886	27	61.607	1	61.343	1
27	29.923	7	29.438	7	59.833	22	59.690	26
28	34.754	1	33.228	1	60.448	11	60.226	15
29	29.132	11	30.155	5	61.094	5	61.075	3
30	28.051	16	27.679	16	60.289	14	60.599	8
31	27.609	22	27.176	21	60.854	7	61.074	4
32	29.972	6	27.482	17	60.799	8	60.728	6

Table 12.11:   Average payoff and rank of the strategies in the final tournament

Strategy	Average payoff	Rank		Rank	Strategy	Average payoff
1	47.131	24		1	25	47.968
2	47.332	21		2	4	47.916
3	47.334	20		3	32	47.881
4	47.916	2		4	28	47.831
5	47.323	22		5	23	47.783
6	47.374	16		6	7	47.757
7	47.757	6		7	21	47.739
8	47.442	12		8	17	47.646
9	47.368	17		9	31	47.640
10	46.759	29		10	16	47.538
11	46.824	28		11	19	47.492
13	46.831	27		12	8	47.442
14	46.260	31		13	29	47.410
15	46.605	30		14	30	47.379
16	47.538	10		15	18	47.375
17	47.646	8		16	6	47.374
18	47.375	15		17	9	47.368
19	47.492	11		18	20	47.350
20	47.350	18		19	22	47.338
21	47.739	7		20	3	47.334
22	47.338	19		21	2	47.332
23	47.783	5		22	5	47.323
24	47.307	23		23	24	47.307
25	47.968	1		24	1	47.131
26	46.834	26		25	27	47.028
27	47.028	25		26	26	46.834
28	47.831	4		27	13	46.831
29	47.410	13		28	11	46.824
30	47.379	14		29	10	46.759
31	47.640	9		30	15	46.605
32	47.881	3		31	14	46.260

The most successful strategy of the final tournament (strategy 25) is also the most simple strategy. The algorithm of this strategy proceeds as follows:

$\alpha=30$ and $\alpha=45$:

-   *If I am the first mover, demand 50 in my first decision step.*
-   *Accept every proposal strictly greater than my alternative value.*
-   *Break off if the opponent's proposal is lower or equal to my alternative value.*

$\alpha=60$ and my alternative$=0$:

-   *If I am the first mover, demand 50 in my first decision step.*
    if I am the second mover and the opponent's proposal is strictly lower than 50, demand 50 in my first decision step.
-   *Accept every proposal greater or equal to 50 in step 2,*
    accept every proposal strictly greater than 0 after step 2.
-   *Break off if the opponent's proposal is equal to 0.*

$\alpha=60$ and my alternative$=60$:

-   *If I am the first mover, demand 80 in my first decision step.*
-   *Accept every proposal strictly greater than 60.*
-   *Break off if the opponent's proposal is lower or equal to 60.*

Except for the case of a low type in a play of a game with $\alpha=60$, the strategy does not make a proposal after the initial one of the first mover. If the strategy is the second mover it accepts every proposal yielding strictly more than the alternative value and it terminates the bargaining by break off in the case of a lower proposal. The initial proposal of the first mover is 50 for $\alpha=30$ and 45, and is 80 for $\alpha=60$. In the case of a low type in a play of a game with $\alpha=60$, the strategy also makes a proposal of 50 as a second mover.

The "individually rational" acceptance level makes this strategy highly exploitable. In particular it is not a best response to itself. But nevertheless, this defensive behavior reached the highest average payoff in the tournament, since it was not exploited by the other strategies, and it did not loose by "non-necessary" break offs. There is no other strategy which is of such a simple form and which has this "individually rational" acceptance level. However, for H_1 in $\alpha=60$ the strategy does not ensure an individually rational payoff for each play.

CHAPTER 13.   TYPICALNESS OF THE FINAL STRATEGIES

In the previous chapter a first analysis of the final strategies was done by studying the average payoffs of the four different types of players which emerged from the strategies. In this chapter we shall investigate the algorithms of the final strategies. It is our aim to draw a picture of the typical strategy. After the introduction of the method of measuring the typicalness of behavior in Section 13.1, we shall apply this method to the final strategies in order to give a picture of the typical strategy (Sections 13.3 to 13.5) and discuss the results of the analysis in Section 13.6. Section 13.2 contains some notes on the evaluation of the typicalness.

13.1   MEASURING THE TYPICALNESS OF BEHAVIOR

This chapter gives a brief description of the method of measuring the typicalness of behavior, which is defined in detail in Kuon (1993). This method is designed to extract the typical behavior from a group of experimental subjects or a set of strategies. For a further application of this method see Kuon (1991).

Suppose the strategies can be described by characteristics which are either present or absent for each strategy. Usually not all characteristics are equally important for the description of the typical strategy. It is desirable to have a method for the determination of weights which express the importance of the characteristics for the description of the typical strategy. These weights will be called *typicities*. Such a method has been proposed in a paper by Selten, Mitzkewitz and Uhlich (1988). The method yields typicities, not only for the characteristics but also for the strategies. The typicity of a strategy expresses the extend to which its algorithm is typical. The definition by Selten, Mitzkewitz and Uhlich is based on an intuitive justification and cannot answer questions concerning the uniqueness and the mathematical properties of the typicities. The entirely different approach by Kuon (1993) of a least squares approximation of the connection between strategies and characteristics by a biproportional matrix leads to the same typicities and allows the formulation of a sufficient condition for the uniqueness of the typicities. It turns out that this condition is always fulfilled if the problem is well specified.

Consider two different ways of describing a structural part of a strategy. It may be described by a feature measurable either on an interval scale or on the boolean scale. We shall refer to qualitative features which are either present or absent as to features measurable on the *boolean scale*. The "value" of a strategy concerning a feature will be called *realization*. In our framework, for example, the realization of the feature "demand of the first mover in the first step" is an integer value between 0 and 100 for each strategy and the feature is thereby measurable on an interval scale, while the realization of the feature "specifies no break off condition" is true or false for each strategy, this means the feature is measurable on the boolean scale.

The question arises under which conditions a strategy should be considered to be typical with respect to a particular feature. To each feature a characteristic is associated which expresses the typical behavior. The characteristic associated to a feature measurable on the boolean scale is specified such that the feature is present in the majority of the observations. A strategy is called *typical with respect to a feature measurable on an interval scale* if its realization is within the interval with the largest occupation surplus among all majority intervals. From each feature a *characteristic* is deduced, such that the characteristic specifies the typical realizations concerning the feature. We say, that *a strategy has a certain characteristic* if its realization is typical with respect to the corresponding feature.

The following will give a concise description of the method. Suppose m strategies are observed. Let X be a feature that is measurable on an interval scale and observable for each strategy. Let $\tilde{x}_j$ be the *realization* of X for strategy $1 \leq j \leq m$. Define a renumbering $\{x_1,...,x_m\} = \{\tilde{x}_1,...,\tilde{x}_m\}$ such that $x_1 \leq x_2 \leq ... \leq x_m$. Then all realizations of X concerning the m strategies lie in the interval $I=[x_1,x_m]$, precisely on $x_1 \leq ... \leq x_m$.

Define the *size of the subinterval* $[x_i,x_j]$ of I as

$$l(x_i,x_j) := \begin{cases} x_j-x_i, & \text{if the realizations can be arbitrary real numbers} \\ x_j-x_i+1, & \text{if the realizations must be discrete values from a finite set} \end{cases}$$

for $1 \leq i \leq j \leq m$.

Define for $l(x_1,x_m) > 0$ and $1 \leq i \leq j \leq m$:

$$L(x_i,x_j) := \frac{l(x_i,x_j)}{l(x_1,x_m)} \quad \textit{relative size} \text{ of } [x_i,x_j]$$

$$p(x_i,x_j) := \#\{x_k \mid 1 \leq k \leq m, \ x_i \leq x_k \leq x_j\} \ \textit{occupation} \text{ of } [x_i,x_j]$$

$$P(x_i,x_j) := \frac{1}{m} p(x_i,x_j) \ \textit{relative occupation} \text{ of } [x_i,x_j]$$

$$S(x_i,x_j) := P(x_i,x_j) - L(x_i,x_j) \quad \textit{occupation surplus} \text{ of } [x_i,x_j]$$

Let $I = [x_1,x_m]$ be the interval of the m realizations of the feature X with $x_1 \leq \ldots \leq x_m$. We say that a subinterval $J \subseteq I$ has the *largest occupation surplus* if

$$S(J) = \max_{1 \leq i \leq j \leq m} S(x_i,x_j).$$

We say that a subinterval $J \subseteq I$ is a *majority interval* if $P(J) > \frac{1}{2}$. We say that a subinterval $M \subseteq I$ has the *largest occupation surplus among all majority intervals* if

$$S(M) = \max_{1 \leq i \leq j \leq m} \left\{ S(x_i,x_j) \mid P(x_i,x_j) > \frac{1}{2} \right\}.$$

An interval with the largest occupation surplus is characterized by the fact that the difference between the relative occupation and the relative size is maximal. For each feature an interval with the largest occupation surplus among all majority intervals will be defined as the interval of the typical behavior and the associated characteristic will express this typical behavior.

In rare cases it may happen that there are two majority intervals M' and M'' with the same occupation surplus. If there is no preference due to the interpretation of the characteristic for one of the intervals, the author would suggest to consider the union of these intervals $M = M' \cup M''$, if it does not coincide with the total interval. In case of a coincidence the formulation of the feature should be changed.

In the above definition only connected intervals are considered. If the realizations come from a set of discrete numbers it is also possible to construct an interval with the largest occupation surplus among all majority intervals as a set of discrete value, if the interpretation of the resulting characteristic is meaningful for the underlying problem.

Let M_i be the (selected) subinterval with the largest occupation surplus among all majority intervals for feature i. We say that *strategy j is typical with respect to the feature i* or *strategy j has the characteristic i* if $x_j \in M_i$. The associated characteristic i is defined as "*the realization of feature i is within M_i*". ($1 \leq i \leq n$, $1 \leq j \leq m$)

If the feature i is measurable on the boolean scale the associated characteristic is specified such that it is present in the majority of the strategies. This means it expresses either the feature itself or the negation of the feature. We say that *strategy j is typical with respect to the feature i* or *strategy j has the characteristic i* if the realization of strategy j concerning feature i is "true". ($1 \leq i \leq n$, $1 \leq j \leq m$)

Now, the *indicator matrix* $A = (a_{ij})_{1 \leq i \leq n, \, 1 \leq j \leq m}$ can be defined with $a_{ij} = 1$ if strategy j has characteristic i and $a_{ij} = 0$ otherwise. This matrix is the basis for the further computations.

The typicities will be defined in such a way that their product is the least squares approximation of the matrix A. This means that the elements of A, the indicators of strategy j having characteristic i, are best approximated by the product of the typicities of characteristic i and strategy j.

Theorem   (Kuon, 1993)
The least squares approximation of A of rank 1 is the matrix $c\gamma s^T$,

$$\text{with } c = \left[\sum_{i=1}^{n} u_i \right]^{-1} \cdot u, \quad s = A^T c, \quad \gamma = \left[\sum_{i=1}^{n} c_i^2 \right]^{-1} \text{ and } u = (u_1, \ldots, u_n) \in \mathbb{R}^n \text{ the orthonormal}$$

eigenvector of AA^T with respect to the largest eigenvalue.

The vector c is called the vector of the *typicities of the characteristics*. The vector s is called the vector of the *typicities of the strategies*. The real number γ is called the *effective number of characteristics*.

Properties of the typicities

(I)   The typicities of the characteristics sum up to 1.

(II) The typicity of a strategy is the sum of the typicities of the strategy's characteristics.

The effective number of characteristics γ can be interpreted as the number of characteristics that are of "relevant influence" for the problem.

Theorem   (Kuon, 1993)

The vectors c and s are uniquely determined, if the matrix AA^T is indecomposable.

This condition is fulfilled if each characteristic is shared by a majority of the strategies. Since this is implied by the construction, the uniqueness of the typicity vectors is ensured.

Remark:

This definition of the typicities differs slightly from the definition in the original approach by Kuon (1993). We chose a simplified presentation of the typicity of a strategy as the sum of the typicities of the strategy's characteristics. In Kuon (1993) this typicity vector is furthermore multiplied by γ. The simplified approach normalizes the typicities of the strategies to the unit interval, and a strategy which has all characteristics has the typicity 1. This simplifies the readability of the evaluations. The value of the effective number of characteristics will be given separately.

13.2   Notes on the Evaluation of the Typicalness

A strategy for the two-person bargaining problem with incomplete information contains the three major specifications of the demand behavior, the acceptance behavior, and break off criteria. These specifications have to be applicable for each value of α, for each of the two types, and for the first and the second mover situation. It is common to all strategies that they give these specifications without an explicit distinction of the first and the second mover situation. The implicit distinction between these situations is due to the fact that the first mover decides in all odd steps and the second mover has to decide in all even steps of the bargaining.

In what follows we shall perform six separate typicity analyses. The three parameter values of α and the two types will be distinguished. No distinction will be made between the two mover situations. Here, the implicit distinction of the strategies will be used. For each of the six problems we shall specify characteristics covering the demand behavior, the acceptance behavior, and the break off specifications. For $\alpha=30$ and $\alpha=45$ this will be expressed by three characteristics. For $\alpha=60$ it seems to be adequate to distinguish between the first demand and the further demands and between the first acceptance specification and the further ones. Therefore, five characteristics are specified in theses cases.

For convenience we shall introduce some ways of speaking. The term *demand* will be used for the demand of a strategy for its own outcome and the term *proposal* will denote the proposal of the opponent for the strategy's outcome. Hence, the proposal is 100 minus the opponent's demand. The term *acceptance level* indicates a lower bound for the acceptance of a proposal. A strategy with an acceptance level accepts every proposal greater or equal to the acceptance level.

All strategies of the experiment specify the acceptance behavior by acceptance levels.

The following three sections contain the formulation of the characteristics and the evaluation of the typicities of the strategies and the characteristics for the three parameter values of α.

13.3   TYPICALNESS OF THE FINAL STRATEGY FOR $\alpha=30$

For both types of players we shall specify three characteristics. They are concerned with the demands, the acceptance prescription and the break off prescription of the strategies. The first two characteristics are identical for the low type and the high type player. In the typical break off prescription the two types are distinguished. While the low type typically does not give a break off prescription, the high type typically breaks off after a predetermined number of steps.

TYPICALNESS OF THE STRATEGY FOR THE LOW TYPE

The first characteristic is concerned with the demands of the strategy.

Characteristic 1:   *The strategy demands 50 in each decision step*

It is common for all strategies that they do not use the history of the bargaining process to determine their demands. 23 of the 31 considered strategies demand 50 in each decision step. The remaining eight strategies change their demands in an only step dependent way. This means they have specified a history independent demand for each decision step. In seven cases (strategies 6, 7, 9, 11, 14, 18, and 30) these fixed lists follow a decreasing order and strategy 21 prescribes an increasing demand sequence. Except for two strategies these lists specify 50 for the first demand of the first mover. In the other two cases it is 65 (for strategy 11) and 63 (for strategy 14), respectively. Three of these fixed lists lower the demand in each decision step by 2 (strategies 7 and 18) and 5 (strategy 9), respectively, such that a revelation of the strategy's type occurs during a sufficiently long bargaining process.

Characteristic 2:   *The strategy specifies an acceptance level of 50 in each decision step*

For 19 strategies the acceptance level is 50 in each step. Strategy 25 (the winning strategy of the final tournament) specifies an acceptance level of 1, and the remaining 11 strategies (5, 6, 7, 9, 14, 16, 18, 19, 20, 30, and 32) prescribe only step dependent decreasing orders of acceptance levels. One of these orders (strategy 32) starts with an acceptance level of 49 and the remaining ten with an acceptance level of 50. It is common for all strategies to consider only the actual proposal for the acceptance decision. No strategy evaluates the proposal process of the opponent in the history of the bargaining process.

Characteristic 3:   *The strategy specifies no break off prescription*

The majority of the strategies (25 out of 31) does not specify a break off prescription. The participants of the strategy experiment knew that a play was terminated by the tournament program (with zero payoffs for both players) at a time where the discounted coalition value was too small to be relevant for the average payoff. Therefore, it was not required to specify a break off prescription in the strategy. A low type player might ultimately hope to reach an agreement and for him a final termination by the tournament program yields the same payoff as a break off. Four strategies (5, 11, 18, and 32) specify a fixed decision step as break off criterion. If the strategy reaches this step it breaks off. Two strategies (20 and 25) condition the break off prescription on the last proposal of the opponent. If the opponent's proposal is lower than a certain bound, the strategy breaks off the bargaining.

There are exactly 17 of the 31 strategies which have all three characteristics. The following two tables present the typicities of the characteristics and the typicities of the strategies.

Table 13.1:   Typicity of the characteristics for the low type of $\alpha=30$

Characteristic	Typicity	#Strategies	Rank
1	.33716	23	2
2	.30701	19	3
3	.35584	25	1

The effective number of characteristics is 2.989.

Table 13.2:   Typicity and average payoff of the strategies for the low type of $\alpha=30$

Strategy	Characteristics			Typicity		Average payoff	
	C1	C2	C3		Rank		Rank
1	•	•	•	1.0000	9	49.967	14
2	•	•	•	1.0000	9	49.967	14
3	•	•	•	1.0000	9	49.967	14
4	•	•	•	1.0000	9	49.967	14
8	•	•	•	1.0000	9	49.967	14
10	•	•	•	1.0000	9	49.967	14
13	•	•	•	1.0000	9	49.967	14
15	•	•	•	1.0000	9	49.967	14
17	•	•	•	1.0000	9	49.967	14
22	•	•	•	1.0000	9	49.967	14
23	•	•	•	1.0000	9	49.967	14
24	•	•	•	1.0000	9	49.967	14
26	•	•	•	1.0000	9	49.967	14
27	•	•	•	1.0000	9	49.967	14
28	•	•	•	1.0000	9	49.967	14
29	•	•	•	1.0000	9	49.967	14
31	•	•	•	1.0000	9	49.967	14
16	•		•	.6930	18.5	49.967	14
19	•		•	.6930	18.5	49.967	14
21		•	•	.6628	20	49.927	28
6			•	.3558	23	49.967	14
7			•	.3558	23	49.935	27
9			•	.3558	23	49.967	14
14			•	.3558	23	49.222	31
30			•	.3558	23	49.967	14
5	•			.3372	27.5	49.967	14
20	•			.3372	27.5	49.648	29
25	•			.3372	27.5	49.550	30
32	•			.3372	27.5	49.967	14
11		•		.3070	30	49.975	1
18				.0000	31	49.967	14

TYPICALNESS OF THE STRATEGY FOR THE HIGH TYPE

The characteristics 1 and 2 coincide with those of the strategies of the low type. However, the majority of the strategies specifies a break off prescription for the high type.

Characteristic 1:   *The strategy demands 50 in each decision step*

All but 5 strategies have this characteristic. The other strategies specify an only step dependent list of demands. Three of theses lists (strategies 18, 21, and 30) start with the equal split as the first demand of the first mover, the remaining two start with 65 (strategy 11) and 61 (strategy 14), respectively. Four of these lists (11, 14, 18, and 30) specify a decreasing demand order and strategy 21 prescribes an increasing order.

Characteristic 2:   *The strategy specifies an acceptance level of 50 in each decision step*

This characteristic is shared by 24 of the 31 strategies. The winning strategy of the final tournament (strategy 25) specifies an acceptance level of 31, and the remaining six strategies (5, 7, 14, 18, 30, and 32) have only step dependent lists of decreasing acceptance levels. Except for strategy 32 these lists start with 50. The first acceptance level of strategy 32 is 49.

Characteristic 3:   *The strategy specifies a break off after a predetermined number of steps*

The majority of the strategies (18 out of 31) specifies a fixed step for the break off. If the strategy reaches this (or a later) decision step it will break off. The distribution of these steps will be shown in table 13.3.

Table 13.3:   Break off steps specified by the strategies

Break off step	Number of strategies
5	3
9	2
10	2
11	2
12	1
16	1
20	1
22	1
30	1
50	3
400	1

Eight strategies (1, 3, 13, 14, 21, 22, 28, and 30) do not specify a break off prescription. The remaining five strategies (5, 7, 15, 25, and 29) specify a break off prescription dependent on the last proposal of the opponent.

There are exactly 15 of the 31 strategies which have all three characteristics. The following two tables present the typicities of the characteristics and the typicities of the strategies.

Table 13.4:   Typicity of the characteristics for the high type of $\alpha = 30$

Characteristic	Typicity	#Strategies	Rank
1	.36699	26	1
2	.35420	24	2
3	.27880	18	3

The effective number of characteristics is 2.960.

<u>Table 13.5:</u>   Typicity and average payoff of the strategies for the high type of $\alpha=30$

Strategy	Characteristics			Typicity		Average payoff	
	C1	C2	C3		Rank		Rank
2	•	•	•	1.0000	8	49.967	14.5
4	•	•	•	1.0000	8	49.967	14.5
6	•	•	•	1.0000	8	49.967	14.5
8	•	•	•	1.0000	8	49.967	14.5
9	•	•	•	1.0000	8	49.967	14.5
10	•	•	•	1.0000	8	49.967	14.5
16	•	•	•	1.0000	8	49.967	14.5
17	•	•	•	1.0000	8	49.967	14.5
19	•	•	•	1.0000	8	49.967	14.5
20	•	•	•	1.0000	8	49.967	14.5
23	•	•	•	1.0000	8	49.967	14.5
24	•	•	•	1.0000	8	49.967	14.5
26	•	•	•	1.0000	8	49.967	14.5
27	•	•	•	1.0000	8	49.967	14.5
31	•	•	•	1.0000	8	49.967	14.5
1	•	•		.7212	19	49.967	14.5
3	•	•		.7212	19	49.967	14.5
13	•	•		.7212	19	49.967	14.5
15	•	•		.7212	19	49.967	14.5
22	•	•		.7212	19	49.967	14.5
28	•	•		.7212	19	49.967	14.5
29	•	•		.7212	19	49.967	14.5
32	•		•	.6458	23	49.967	14.5
11		•	•	.6330	24	49.975	1
5	•			.3670	26	49.967	14.5
7	•			.3670	26	49.648	29
25	•			.3670	26	49.550	30
21		•		.3542	28	49.927	28
18			•	.2788	29	49.967	14.5
14				.0000	30.5	49.434	31
30				.0000	30.5	49.967	14.5

13.4   TYPICALNESS OF THE FINAL STRATEGY FOR $\alpha=45$

Like for the strategies for plays of games with $\alpha=30$ we shall introduce three characteristics, describing the demands, the acceptance prescription and the break off prescription
of a strategy. The characteristics of the strategies for the high type coincide with those
introduced for $\alpha=30$, the characteristics for the strategies of the low type differ in the
characteristic describing the acceptance prescription from the case of $\alpha=30$.

TYPICALNESS OF THE STRATEGY FOR THE LOW TYPE

Characteristic 1:   *The strategy demands 50 in each decision step*

This characteristic is shared by 19 of the 31 strategies. Ten (strategies 1, 6, 7, 11, 14, 18,
27, 28, 29, and 30) of the twelve remaining strategies lower the demand in an only step
dependent way. Four of these strategies (6, 7, 18, and 30) start with an initial demand of 50,
and the other ones start with 54, 55, 60, 63, or 72. Strategy 21 increases its demand by a
step dependent list. Strategy 9 lowers its demand dependent on the opponent's behavior. If
the opponent's last proposal is a repetition of the previous proposal he is classified as "not
flexible" by the strategy. In this case the strategy lowers its demand starting form a demand
of 45 in decision step 30 by 5 in each following decision step. In the other case the strategy
remains at a demand of 50. Three of the strategies which specify a decreasing demand order
do this by lowering by a fixed number (2 for strategies 7 and 18, and 5 for strategy 9,
respectively), such that the strategy reveals its type by proposing less than 45 during the
bargaining process.

Characteristic 2:   *The strategy has an acceptance level of 50 in each decision step or it*
                         has an acceptance level of 50 in the first decision step of the second
                         mover and the further acceptance levels decrease in an only step
                         dependent order

14 strategies (2, 3, 4, 8, 10, 11, 13, 15, 17, 21, 22, 23, 24, and 32) prescribe an acceptance
level of 50 in each decision step, and 11 strategies (5, 6, 7, 14, 16, 18, 19, 26, 27, 28, and

30) specify an acceptance level of 50 in the first decision step of the second mover and then lower the acceptance level in an only step dependent way. This means that 25 strategies have the above characteristic. Two strategies prescribe an acceptance level of 49 in each decision step (20 and 29). Two strategies specify acceptance levels lower than 49 (strategy 25 specifies 1 and strategy 31 specifies 46, respectively). Strategy 1 specifies an acceptance level of 55. The remaining strategy 9 is the only one which specifies an acceptance level dependent on the history of the bargaining process. If the opponent is classified as "not flexible" (see above) the acceptance level is lowered by 5 starting from 45 in decision step 30. If the opponent is classified as "flexible" the acceptance level is 50.

Characteristic 3:   *The strategy specifies no break off prescription*

This characteristic is present in 26 of the 31 strategies. Four strategies (5, 11, 18, and 32) specify a fixed step for the break off and strategy 25 conditions the break off prescription on the last proposal of the opponent. If the opponent proposed zero, the strategy breaks off.

There are exactly 15 of the 31 strategies which have all three characteristics. The following two tables present the typicities of the characteristics and the typicities of the strategies.

Table 13.6:   Typicity of the characteristics for the low type of $\alpha=45$

Characteristic	Typicity	#Strategies	Rank
1	.28073	19	3
2	.35645	25	2
3	.36282	26	1

The effective number of characteristics is 2.963.

Table 13.7: Typicity and average payoff of the strategies for the low type of $\alpha=45$

Strategy	Characteristics			Typicity		Average payoff	
	C1	C2	C3		Rank		Rank
2	•	•	•	1.0000	8	46.592	24
3	•	•	•	1.0000	8	46.592	24
4	•	•	•	1.0000	8	46.592	24
8	•	•	•	1.0000	8	46.592	24
10	•	•	•	1.0000	8	46.592	24
13	•	•	•	1.0000	8	46.592	24
15	•	•	•	1.0000	8	46.592	24
16	•	•	•	1.0000	8	47.242	13.5
17	•	•	•	1.0000	8	46.596	18
19	•	•	•	1.0000	8	47.242	13.5
20	•	•	•	1.0000	8	47.364	12
22	•	•	•	1.0000	8	46.592	24
23	•	•	•	1.0000	8	46.592	24
24	•	•	•	1.0000	8	46.592	24
26	•	•	•	1.0000	8	48.845	1
6		•	•	.7193	19	47.559	10
7		•	•	.7193	19	47.976	8
14		•	•	.7193	19	44.723	31
21		•	•	.7193	19	47.777	9
27		•	•	.7193	19	48.197	5
28		•	•	.7193	19	47.392	11
30		•	•	.7193	19	46.592	24
31	•		•	.6436	23	48.059	6
5	•	•		.6372	24	48.760	3
1			•	.3628	26	45.746	30
9			•	.3628	26	47.139	15
29			•	.3628	26	48.009	7
11		•		.3565	28.5	46.868	17
18		•		.3565	28.5	46.893	16
25	•			.2807	30.5	48.377	4
32	•			.2807	30.5	48.764	2

TYPICALNESS OF THE STRATEGY FOR THE HIGH TYPE

Characteristic 1:   *The strategy demands 50 in each decision step*

18 of the 31 strategies have this characteristic. From the remaining 13 strategies, strategy 26 specifies a constant demand of 65 for all decision steps. Eight sequences (for strategies 1, 11, 14, 17, 18, 21, 27, and 30) are only step dependent and start with initial demands of 50, 54, 62, 70, or 72. Strategy 21 specifies a step dependent increasing order of demands, while the other strategies specify decreasing sequences of demands. Four sequences (strategies 9, 28, 29, and 32) depend on the proposals of the opponent. They start with initial demands of 54, 55, or 60. Strategy 32 conditions on the fact whether the opponent's last proposal is higher, equal or lower than the opponent's proposal before the last proposal. Strategy 28 conditions on the question whether the opponent's last proposal was greater or smaller than 50. Strategy 29 conditions on the question whether the opponent revealed with his last proposal, and the strategy 9 conditions on its classification of the opponent. It classifies an opponent as "not flexible" if he repeated his last proposal.

Characteristic 2:   *The strategy specifies an acceptance level of 50 in each decision step*

This characteristic is shared by 19 strategies. Seven strategies also have a constant acceptance level implemented in their strategy, but it is different from 50. These acceptance levels are: 46 (for strategies 31 and 25), 49 (for strategy 32), 51 (for strategy 1), 53 (for strategy 28), 55 (for strategy 29), and 60 (for strategy 26). Strategy 27 has an increasing acceptance level, starting from 51, and strategies 5, 7, 18, and 14 have decreasing acceptance levels starting from 50. All these orders are only step dependent.

Characteristic 3:   *The strategy specifies a break off after a predetermined number of*
                      steps

Three strategies (3, 21, and 22) do not specify a criterion for break off. Six strategies specify a break off criterion conditional on the last proposal of the opponent (7, 8, 9, 14, 15, and 25). The remaining 22 strategies condition the break off on a predetermined step in the bargaining. If this step is reached, the strategy breaks off. The distribution of these break off steps is shown in table 13.8.

Table 13.8:   Break off steps specified by the strategies

Break off step	Number of strategies
2	1
4	2
5	3
6	3
7	4
8	1
9	1
10	1
11	1
12	2
13	1
16	1
25	1

There are exactly 10 of the 31 strategies which have all three characteristics. The following two tables present the typicities of the characteristics and the typicities of the strategies.

Table 13.9:   Typicity of the characteristics for the high type of $\alpha=45$

Characteristic	Typicity	#Strategies	Rank
1	.31911	18	3
2	.33543	19	2
3	.34546	22	1

The effective number of characteristics is 2.997.

Table 13.10: Typicity and average payoff of the strategies for the high type of $\alpha=45$

Strategy	Characteristics			Typicity		Average payoff	
	C1	C2	C3		Rank		Rank
2	•	•	•	1.0000	5.5	49.488	3.5
4	•	•	•	1.0000	5.5	49.413	12
6	•	•	•	1.0000	5.5	49.441	9.5
10	•	•	•	1.0000	5.5	49.463	6.5
13	•	•	•	1.0000	5.5	49.403	13
16	•	•	•	1.0000	5.5	49.389	15.5
19	•	•	•	1.0000	5.5	49.474	5
20	•	•	•	1.0000	5.5	49.426	11
23	•	•	•	1.0000	5.5	49.488	3.5
24	•	•	•	1.0000	5.5	49.389	15.5
11		•	•	.6809	12	49.449	8
17		•	•	.6809	12	49.001	26
30		•	•	.6809	12	49.441	9.5
5	•		•	.6646	14.5	49.120	22
31	•		•	.6646	14.5	49.493	2
3	•	•		.6545	17.5	49.389	15.5
8	•	•		.6545	17.5	49.110	23
15	•	•		.6545	17.5	49.463	6.5
22	•	•		.6545	17.5	49.389	15.5
1			•	.3455	23	49.096	25
18			•	.3455	23	49.381	18
26			•	.3455	23	44.567	31
27			•	.3455	23	44.594	30
28			•	.3455	23	45.334	29
29			•	.3455	23	45.789	28
32			•	.3455	23	49.098	24
9		•		.3354	27.5	49.345	19
21		•		.3354	27.5	49.568	1
7	•			.3191	29.5	49.329	20
25	•			.3191	29.5	49.160	21
14				.0000	31	48.663	27

13.5   TYPICALNESS OF THE FINAL STRATEGY FOR $\alpha=60$

The strategies for the plays of games with $\alpha=60$ will be described be five characteristics. The demand as well as the acceptance prescription will each be covered by two characteristics, one focusing on the first action and the other one focusing of the development of the prescription. The characteristic concerning the initial demand of the first mover and the characteristic concerning the initial acceptance level of the second mover will be specified as a subinterval with the largest occupation surplus among all majority intervals of the respective distribution (see Section 13.1).

TYPICALNESS OF THE STRATEGY FOR THE LOW TYPE

Characteristic 1:   *The initial demand of the first mover is either 70 or 80.*

The distribution of the initial demands of the first mover is given in the following table.

Table 13.11:   Distribution of the initial demand of the first mover

Initial demand	Number of strategies
40	1
50	7
55	1
60	1
63	1
65	1
70	8
75	2
79	1
80	8

Characteristic 1 is present in 16 strategies.

Characteristic 2:   *The further demands decrease in an only step dependent order*

16 strategies prescribe a decreasing lists of demands which only depends on the bargaining step (strategies 1, 2, 4, 5, 6, 13, 14, 18, 20, 22, 24, 26, 27, 29, 30, and 31). Five strategies specify a demand sequence which is identical for all steps (strategies 8, 10, 11, 15, and 25). The remaining 10 strategies specify their demands in an opponent dependent way. This mostly happens by classifying the opponent's last proposal into different ranges and reacting with different demands. As an example the classification of strategy 19 shall be reported since it is very much in the style the other strategies classify. If the opponent's last proposal is smaller or equal to 20, the strategy demands 70, if it is strictly greater than 20 but strictly smaller than 50, the strategy demands 50. A proposal greater or equal to 50 is accepted by the strategy. The classification can be seen as an attempt to identify the opponent's "toughness". If the opponent's last proposal is in a "low" range, the strategy demands a high value. This means it repeats high demands with high demands. If, on the other hand, the opponent's last proposal is in a range around the equal split, the strategy demands the equal split. A proposal greater than the equal split is accepted.

All strategies, except strategies 5, 11, 15, and 21, reveal their type during the demand process, but most of the strategies do not make demands which leave individually rational amounts for the strong player. They reach an agreement with the strong player in the "passive" way of accepting. Only 10 strategies (3, 6, 7, 15, 17, 22, 27, 28, 29, and 32) make demands which leave more than 60 to the opponent. This may either happen by a step dependent decrease of the demands or after the strategy classified the opponent as strong.

Characteristic 3:   *The initial acceptance level of the second mover is 50*

The distribution of the initial acceptance levels of the second mover is given in the following table.

Table 13.12:   Distribution of the initial acceptance level of the second mover

Initial acceptance level	Number of strategies
0	1
25	1
30	2
40	2
50	16
59	1
60	3
65	1
70	1
80	3

Characteristic 3 is present in 16 strategies.

Characteristic 4:   *The further acceptance levels are either identical to the initial one or decrease in an only step dependent order*

15 strategies prescribe acceptance levels which are decreasing in a step dependent way (strategies 1, 2, 3, 4, 5, 6, 8, 14, 16, 19, 20, 22, 23, 27, and 30). Nine strategies specify an acceptance level which is identical for all steps (strategies 10, 11, 13, 15, 18, 21, 24, 25, and 31). Therefore 24 strategies fulfill characteristic 4. The seven remaining strategies (7, 9, 17, 26, 28, 29, and 32) make their acceptance level dependent on the opponent's proposals during the bargaining process.

Two different ways to prescribe an opponent dependent acceptance level can be observed in the strategies. One way is a classification of the opponent's last proposal into predetermined ranges. This procedure works in an analogous way as described for the opponent dependent demands. The strategies 7, 17, and 28 use these classification dependent adaptations. The second way is to evaluate the opponent's proposal process over the bargaining process. Here, it is evaluated whether the opponent decreased his proposals during the bargaining process. A decrease in the opponent's proposal is seen as a severe threat and the own acceptance level is lowered in this case. This procedure is applied in strategy 9, 26, and 32. Strategy 29 implemented a combination of both procedures.

Characteristic 5:   *The strategy specifies no break off prescription*

This characteristic is shared by 27 strategies. Three strategies prescribe a break off after a fixed number of steps (strategies 11, 18, and 20), and strategy 25 makes the break off decision dependent on the opponent's last proposal. If this was zero, then the strategy breaks off.

There are exactly 4 of the 31 strategies which have all five characteristics. The following two tables present the typicities of the characteristics and the typicities of the strategies.

Table 13.13:   Typicity of the characteristics for the low type of α=60

Characteristic	Typicity	#Strategies	Rank
1	.17062	16	3
2	.16915	16	4
3	.16813	16	5
4	.23816	24	2
5	.25394	27	1

The effective number of characteristics is 4.826.

Table 13.14:   Typicity and average payoff of the strategies for the low type of $\alpha = 60$

Strategy	Characteristics					Typicity		Average payoff	
	C1	C2	C3	C4	C5		Rank		Rank
1	•	•	•	•	•	1.0000	2.5	27.966	16
22	•	•	•	•	•	1.0000	2.5	27.581	21
24	•	•	•	•	•	1.0000	2.5	27.070	23
27	•	•	•	•	•	1.0000	2.5	29.680	6
2	•	•		•	•	0.8319	6.5	27.975	15
4	•	•		•	•	0.8319	6.5	30.306	4
5	•	•		•	•	0.8319	6.5	26.857	26
31	•	•		•	•	0.8319	6.5	27.392	22
16	•		•	•	•	0.8309	10	28.912	10
19	•		•	•	•	0.8309	10	28.615	13
23	•		•	•	•	0.8309	10	30.773	3
6		•	•	•	•	0.8294	12.5	27.710	19
13		•	•	•	•	0.8294	12.5	25.059	29
20	•	•	•	•		0.7461	14	26.995	24
14		•		•	•	0.6613	15.5	27.706	20
30		•		•	•	0.6613	15.5	27.865	17
8			•	•	•	0.6602	17.5	28.749	11
10			•	•	•	0.6602	17.5	24.732	30
28	•		•		•	0.5927	19	33.991	1
26		•	•		•	0.5912	20	26.184	27
11	•		•	•		0.5769	21	25.077	28
3			•		•	0.4921	23	26.976	25
15			•		•	0.4921	23	23.471	31
21			•		•	0.4921	23	29.234	9
9	•				•	0.4246	25.5	27.790	18
17	•				•	0.4246	25.5	29.754	5
29		•			•	0.4231	27	29.643	7
7			•		•	0.4221	28	29.407	8
18		•		•		0.4073	29	28.250	14
32					•	0.2539	30	28.727	12
25				•		0.2382	31	31.424	2

TYPICALNESS OF THE STRATEGY FOR THE HIGH TYPE

<u>Characteristic 1:</u>   *The initial demand of the first mover is either 70 or 80.*

The distribution of the initial demands of the first mover is given in table 13.15.

<u>Table 13.15:</u>   Distribution of the initial demand of the first mover

Initial demand	Number of strategies
70	6
75	4
79	2
80	14
85	1
immediate break off	4

Therefore 20 strategies have characteristic 1. It is interesting to observe that there are four strategies which immediately break off without making any proposal.

<u>Characteristic 2:</u>   *The further demands are either identical to the initial demand or decrease in an only step dependent order*

Twelve strategies specify a demand sequence which is identical for all steps (strategies 1, 6, 8, 10, 11, 15, 16, 18, 19, 22, 25, and 27). Ten strategies prescribe a decreasing list of demands which only depends on the bargaining step (strategies 3, 4, 5, 7, 14, 20, 26, 28, 30, and 31). In these lists the lowest demand is between 60 and 80, mostly at 70. This means that the above characteristic is fulfilled by 22 strategies. Two of the remaining strategies specify an increasing sequence of demands. The demands increase to 95 (for strategy 23) and to 80 (for strategy 24). Three strategies (17, 29, and 32) make their demands contingent on the opponent's proposals during the bargaining process. Four strategies (2, 9, 13, and 21) do not prescribe any demand for the case of a high type player in a play of $\alpha=60$. The strategies either accept or break off in each decision step.

Characteristic 3:   *The initial acceptance level of the second mover is between 60 and 70*
                        (boundaries included).

The distribution of the initial acceptance levels of the second mover is given in the following
table.

Table 13.16:   Distribution of the initial acceptance level of the second mover

Initial acceptance level	Number of strategies
60	5
61	6
65	4
70	7
71	1
75	3
79	1
80	3

22 strategies have characteristic 3. Strategy 21 does not specify an acceptance level. Independent of the proposal of the opponent it breaks off the bargaining in each decision step.

Characteristic 4:   *The further acceptance levels are identical to the initial acceptance*
                        level

27 of the 31 strategies have this characteristic. The strategies 4 and 30 specify a decreasing
acceptance level which only depends on the bargaining step, strategy 17 specifies an acceptance level conditioning on the opponent's last proposal, and strategy 21 does not specify any
acceptance level.

Characteristic 5:   *The strategy breaks off if the opponent's proposal is lower than a*
                        predetermined bound

This characteristic is shared by 18 strategies. Mostly, the predetermined bound is 60. Four of these strategies (1, 23, 26, and 28) specify in addition an upper bound for the number of bargaining steps. If this step is reached the bargaining is terminated by break off any way. The remaining 13 strategies prescribe a break off after a fixed number of steps (strategies 3, 5, 6, 8, 10, 11, 18, 20, 21, 22, 24, 27, and 31).

There are exactly 6 of the 31 strategies which have all five characteristics. The following two tables present the typicities of the characteristics and the typicities of the strategies.

Table 13.17:   Typicity of the characteristics for the high type of $\alpha=60$

Characteristic	Typicity	#Strategies	Rank
1	.19388	20	4
2	.20660	22	2
3	.20608	22	3
4	.24059	27	1
5	.15285	18	5

The effective number of characteristics is 4.902.

Table 13.18:   Typicity and average payoff of the strategies for the high type of $\alpha = 60$

Strategy	Characteristics					Typicity		Average payoff	
	C1	C2	C3	C4	C5		Rank		Rank
1	•	•	•	•	•	1.0000	3.5	60.046	16
7	•	•	•	•	•	1.0000	3.5	60.247	14
14	•	•	•	•	•	1.0000	3.5	57.810	31
16	•	•	•	•	•	1.0000	3.5	59.750	25.5
19	•	•	•	•	•	1.0000	3.5	59.689	27
25	•	•	•	•	•	1.0000	3.5	59.750	25.5
3	•	•	•	•		.8472	10.5	61.115	3
5	•	•	•	•		.8472	10.5	59.267	30
6	•	•	•	•		.8472	10.5	59.597	28.5
8	•	•	•	•		.8472	10.5	60.265	13
10	•	•	•	•		.8472	10.5	59.831	22
20	•	•	•	•		.8472	10.5	60.700	8
22	•	•	•	•		.8472	10.5	60.531	10
31	•	•	•	•		.8472	10.5	60.964	5
26	•	•		•	•	.7939	15	61.475	1
27		•	•	•		.6533	16	59.761	24
11	•	•		•		.6411	17	59.597	28.5
24	•		•	•		.6406	18	60.859	6
15		•		•	•	.6000	19.5	60.170	15
28		•		•	•	.6000	19.5	60.337	12
2			•	•	•	.5995	22.5	60.000	18.5
9			•	•	•	.5995	22.5	60.000	18.5
13			•	•	•	.5995	22.5	60.000	18.5
32			•	•	•	.5995	22.5	60.763	7
23	•			•	•	.5873	25	59.913	21
30		•	•		•	.5655	26	60.444	11
4	•	•			•	.5533	27	61.253	2
17	•		•		•	.5528	28	60.588	9
18		•		•		.4472	29	59.790	23
29				•	•	.3934	30	61.084	4
21						.0000	31	60.000	18.5

13.6 A PICTURE OF THE TYPICAL STRATEGY

A strategy for the two-person bargaining game with incomplete information has to contain three major parts. It has to state a demand, an acceptance prescription and a break off prescription for every history. In Sections 13.3 to 13.5 we introduced characteristics of the strategies covering these three parts which are present in the majority of all strategies. Together these typical characteristics give a picture of the typical strategy of the strategy experiment. The following table gives an overview of the characteristics.

Table 13.19:   Characteristics of the strategies

α	Type	Demands	Acceptance level	Break off
30	LO	50 in each step	50 in each step	no break off
	HI	50 in each step	50 in each step	break off after a predetermined step
45	LO	50 in each step	50 in each step or 50 in the first step and the further acceptance levels decrease in an only step dependent order	no break off
	HI	50 in each step	50 in each step	break off after a predetermined step
60	LO	1. step: 70 or 80 and the further demands decrease in an only step dependent order	2. step: 50 and the further acceptance levels are identical to the initial one or decrease only step dependent	no break off
	HI	1. step: 70 or 80 and the further demands are identical to the initial one or decrease in an only step dependent order	2. step: 60,...,70 and the further acceptance levels are identical to the initial one	break off dependent on opponent's proposal

Note, that we do not distinguish between the first and the second mover. The fact that, especially for $\alpha=30$ and $\alpha=45$, the majority of strategies specifies the same action in each step shows that the strategies do not distinguish between the first and the second mover situation. In case the strategies do not specify identical actions in each step, they mostly specify step contingent monotonic orders, which distinguish between first mover and second mover, but only in the implicit way that the first mover decides at all odd decision steps and the second mover at all even ones. The structure of the behavior of the two mover types is not different. Therefore, we refrained from distinguishing between the first and the second mover.

In this section we shall discuss the characteristics and correlate the typicity of a strategy with its success in the situation it was designed for. For this purpose a Spearman rank correlation analysis will be conducted between the typicity and the average payoff of a strategy for each of the six situations. This means, for example, we shall correlate the typicity of a strategy for the low type of $\alpha=45$ with the success of this strategy in plays of the low type of $\alpha=45$. The average payoff and the rank of the strategy resulting from this payoff was already given in the Sections 13.3 to 13.5 in the tables displaying the typicities of the strategies.

The most remarkable result of the analysis is that the strategies very rarely react to the proposals of the opponent, but instead prescribe an action which is identical for all bargaining histories or only depends on the actual bargaining step. For $\alpha=30$ no strategy makes its demands dependent on past proposals of the opponent. A demand which conditions on the opponent's past proposals is observed four times for $\alpha=45$ and 11 times for $\alpha=60$, but in both cases this is not typical. An analogous result was found for the acceptance prescription. This prescription was always specified in the form of an acceptance level. This means a lower bound was specified and each proposal which was greater than this bound was accepted. For $\alpha=30$ the acceptance levels were either constant for every history or only step dependent. No strategy conditioned the acceptance level on the past proposals of the opponent. For $\alpha=45$ we observe only one strategy and for $\alpha=60$ we find seven strategies which make the acceptance level dependent on the opponent's past proposals. In none of the cases an opponent dependent adaptation of the acceptance level is part of a typical strategy. The break off decision, which is typically only present in strategies for the high type, depends for $\alpha=30$ and $\alpha=45$ typically only on the bargaining step. For $\alpha=30$ we observe four strategies and for $\alpha=45$ we observe six strategies which condition the break off decision on the opponent's proposal. For the high type of $\alpha=60$ it is typical to evaluate the opponent's proposals for the break off decision. Break off prescriptions, not only considering the opponent's actual proposal, but the complete past proposal process, can be found in 18 strategies.

In the seminar paper which had to explain the development and the reasoning of the strategy some participants expressed that an observation of the opponent's bargaining process was not worthwhile, since most bargaining processes are too short to allow valid conclusions.

Very rarely the demand or acceptance orders are decreasing regularly in the step number. Mostly, they are constant for several steps.

Furthermore, it should be remarked that no strategy uses random moves.

TYPICITIES OF THE STRATEGIES FOR $\alpha=30$

Together the three characteristics for the low type describe a complete strategy. This very simple strategy was programmed by 17 of the 31 participants. A population of players all playing this strategy ends up in immediate equal split agreements, which was observed in most of the plays of the introductory game playing sessions and in the plays of experienced players in the game playing experiment, reported in Part II. The one-sided Spearman rank correlation analysis shows that the typicity of the strategy is positively correlated to the success of this strategy at a significance level of .1, which means that the typical strategies are also the successful ones.

For the high type the characteristics specify the same demand and acceptance prescription, but it is typical to break off the bargaining after a predetermined number of steps. The distribution of these steps was shown in table 13.3 in Section 13.3. Mostly, a step between step 5 and step 11 was chosen. Also for the high type strategy the one-sided Spearman rank correlation analysis finds a positive correlation between the typicity and the success of the strategy at a significance level of .01.

TYPICITIES OF THE STRATEGIES FOR $\alpha=45$

The characteristics for the low type formulating the demands and the break off pre-scription are identical to the ones for $\alpha=30$. The strategy demands 50 in each decision step and never breaks off. The typical acceptance prescription specifies an acceptance level of 50 in each step or an acceptance level of 50 in the first step and an only step dependent decreas-ing sequence of further acceptance levels. The one-sided Spearman rank correlation analysis finds a correlation between the typicity and the success of the strategy, but a negative one (at a significance level of .01). This means that here the typical strategies are the ones with the low average payoffs. Where does this come from? Table 13.7 shows that most of the strategies with typicity 1 have a payoff rank of 19. A closer look shows that these are the strategies with an acceptance level of 50 in each decision step. Strategy 26, for example, which has the highest average payoff, also has the typicity 1, but here characteristic 2 is present in the from of a decreasing acceptance level. The strategy with the second best average payoff even specifies an acceptance level of 49 in the first step and from then on a decreasing order. A decreasing order of acceptance levels or a constant acceptance level lower than 50 (like strategies 25 and 31) leads to higher average payoffs than a constant acceptance level of 50. However, the demand that each characteristic has to be present in the majority of the strategies makes it necessary to include the feature of a constant acceptance level of 50 into the characteristic, since it is present in 14 strategies. This is the feature with the highest occupation among all strategies, but it does not lead to a successful performance.

For the high type, we find the same typical characteristics as for $\alpha=30$. A strategy typically demands 50 in each step, accepts every proposal greater than 50 in each step and breaks off after a predetermined number of steps, mostly between 4 and 7. The one-sided Spearman rank correlation analysis finds a positive correlation between the typicity and the success of the strategy at a significance level of .01. Here again, the typical players are also the successful ones. The "non-typical" players are, for example, those which start with high initial demands between 55 and 70 and lower these demands very slowly. These are the strategies which make an acceptance level of constantly 50 so much unfavorable for the low type player. These plays then mostly end up in break offs or in equal split agreements after a large number of steps, which both implies a reduction of the average payoff.

TYPICITIES OF THE STRATEGIES FOR $\alpha = 60$

It is typical for the strategies of the high type as well as for strategies for the low type to demand 70 or 80 in the first decision step of the first mover. This means it is typical for the low type to imitate the high type in the first demand. However, the two types are distinguished with respect to the other characteristics. The further demands of the low type shrink in an only step dependent way, while it is typical for the high type to repeat his demands. Although, the low type starts with high initial demands the typical strategy accepts a proposal of 50 in the first decision step of the second mover and typically remains at this acceptance level or lowers it in an only step dependent way during the bargaining process. In contrast, a high type player has an initial acceptance level between 60 and 70 and typically does not deviate from this level. This shows that typically the high types are satisfied with only a small additional gain. The typical proposals accepted are lower than the split of the surplus in addition to the high alternative value (which is 80 in this case).

For the strategies of $\alpha = 30$ and $\alpha = 45$ the typical acceptance level was equal to the typical demand. For $\alpha = 60$ the acceptance levels are smaller than the demands. For the low type this is explained by the attempt to reach a high agreement by bluffing with a high demand, but on the other hand a lower acceptance level lowers the risk of break offs. The high type player typically makes moderate demands, and is willing to accept even more moderate proposals. Like in the game playing experiment the high type is not able to achieve payoffs as high as in comparable games under complete information (see Kuon et al. 1993), since he is not able to "prove" his strength.

Like for the other parameter values of α, the low type player typically does not initiate a break off. The high type strategy conditions the break off prescription on the proposal of the opponent.

The Spearman rank correlation analysis cannot find a correlation between the typicity and the success of the strategy of a low type at a significance level of .2 (two-sided). This means that the typical strategies, which can be described as "bluffing" by high demands in the first decision step of the first mover, but being much more moderate in the acceptance behavior and the further demands, are not significantly correlated to the successful ones. For the strategy of the high type the one-sided Spearman rank correlation analysis finds a negative correlation between the typicity and the success of the strategy at a significance level of .05. This means, that the typical strategies with the more moderate demands and acceptance levels are the less successful ones. By looking at the successful strategies we want to study the differences to the typical ones in order to understand why the typical strategies are less successful. The two strategies which are most successful (26 and 4) have higher initial demands which are kept constant or decrease, respectively. The strategies with payoff ranks 3, 5, 6, and 8 share all characteristics but the last one. They specify only step dependent break offs. The strategies with payoff ranks 4 and 7 specify their demands in an opponent dependent way. However, no clear picture emerges, why the "untypical" strategies are more successful. There seem to be various alternative ways to gain higher payoffs than the typical strategies. But, they do not hint at a precise misconception of the typical strategy, like this could be found for the low type of $\alpha=45$.

Table 13.20 summarizes the results of the Spearman rank correlation analysis.

Table 13.20:   Results of the one-sided Spearman rank correlation test between the typicity and the success of a strategy

α	Type	Correlation	Significance level
30	LO	positive	.10
	HI	positive	.01
45	LO	negative	.01
	HI	positive	.01
60	LO	not significant at .2 (two-sided)	
	HI	negative	.05

The effective number of characteristics is in all cases close to the total number of characteristics, which shows that all characteristics are of relevance for the underlying problem. The typicities of the characteristics do not show significant differences, which means that the characteristics are equally important for the picture of the typical strategy.

CHAPTER 14.   AN EVOLUTIONARY TOURNAMENT

Consider a world of individuals being involved in the two-person bargaining game with incomplete information. The world consists of four different populations: the weak first movers, the strong first movers, the weak second movers, and the strong second movers. A member of a first mover population interacts with the members of both second mover populations and vice versa. The four populations act according to the strategies of the strategy experiment. Suppose, in the beginning of the world every strategy has the same number of representatives in each of the four populations. The number of representatives of a strategy in a population in the following period depends on its fitness. One could pose the question how the four populations develop under this dynamic and whether the evolutionary selection by fitness converges to an equilibrium of the one-shot game.

This informally described evolutionary approach will be introduced in a concise way in Section 14.1. Section 14.2 will present the results of the evolutionary tournament and the last section will discuss the connection with the typicity analysis of Chapter 13.

14.1   THE EVOLUTIONARY APPROACH

Let α be arbitrary but fixed, and let $N := \{1,\dots,11,13,\dots 32\}$. Consider four large populations p_i. A member of a population p_i plays a strategy from the set S_i ($i = 1,\dots,4$):

$S_1 = \{L_1(i) \mid L_1(i)$ is the strategy for the low type first mover of participant i, $i \in N\}$,
$S_2 = \{H_1(i) \mid H_1(i)$ is the strategy for the high type first mover of participant i, $i \in N\}$,
$S_3 = \{L_2(i) \mid L_2(i)$ is the strategy for the low type second mover of participant i, $i \in N\}$,
$S_4 = \{H_2(i) \mid H_2(i)$ is the strategy for the high type second mover of participant i, $i \in N\}$.

The number of a strategy refers to the number in the strategy experiment.

Assume that each member of population p_1 meets with equal probability a member of population p_3 and a member of population p_4. The same is true for the members of population p_2. Similarly, a member of population p_3 meets with equal probability a member of population p_1 and a member of population p_2, and the same is true for the members of population p_4. Meeting a member of another population means playing the two-person bargaining game with incomplete information according to the strategies of the two members which meet.

Let $a(i,j;p_k,p_m)$ denote the payoff of an individual of population p_k playing strategy i in a play with an individual of population p_m playing strategy j, $i,j \in N$, and either $p_k=1,2$ and $p_m=3,4$ or $p_k=3,4$ and $p_m=1,2$. Note that $a(j,i;p_m,p_k)$ denotes the payoff of the opponent of the above play. (The payoff of an individual of population p_k playing strategy i will shortly be called the payoff of strategy i of population p_k.)

At the beginning of the world each strategy has the same *weight* (representation) in the respective population. During the evolution of the world of the four populations the representation of a strategy changes according to its fitness. The fitness of a strategy in period t or in other words the expected number of offsprings in period t+1 is the average payoff in period t. This evolutionary approach is expressed in the *discrete replicator dynamics*, which takes the payoff of a strategy relative to the payoffs of the other strategies of the same population as a measure of fitness.

Observe the evolutionary process at the discrete time $t=0,1,2,...$, where $t=0$ is the beginning of the process. Let $w_t(i,p_k)$ denote the *weight* of strategy i in population p_k at time t, $i \in N$, and $k=1,...,4$, $t \geq 0$. At the beginning of the evolutionary process all strategies of all populations have equal weights.

$$w_0(i,p_k) = \frac{1}{31} \text{ for all } k=1,..,4 \text{ and } i \in N.$$

Let $\pi_t(i,p_k)$ denote the *average payoff (fitness)* of strategy i of population p_k in time t ($i \in N$ and $t \geq 0$).

$$\pi_t(i,p_k) = \frac{1}{2} \sum_{j \in N} [w_t(j,p_3)a(i,j;p_k,p_3) + w_t(j,p_4)a(i,j;p_k,p_4)], \quad k=1,2,$$

$$\pi_t(i,p_k) = \frac{1}{2} \sum_{j \in N} [w_t(j,p_1)a(i,j;p_k,p_1) + w_t(j,p_2)a(i,j;p_k,p_2)], \quad k=3,4.$$

The *discrete replicator dynamics* determines the weight in time t+1 as:

$$w_{t+1}(i,p_k) = (1-31\epsilon) \frac{w_t(i,p_k)\pi_t(i,p_k)}{\sum\limits_{j\in N} w_t(j,p_k)\pi_t(j,p_k)} + \epsilon, \quad t\geq 0, \ i\in N, \ \text{and} \ k=1,\ldots,4.$$

The weight of a strategy in period t+1 depends on its payoff in the population weighted with the relative frequency of representatives playing this strategy in period t, divided by the weighted sum of the payoffs of all strategies of the population. By this dynamic equation the weight of each strategy evolves over time contingent on the fitness and the weights of all strategies.

The variable ϵ has to be interpreted as the probability of an emergence of a mutant in the population. Only mutants which emerge in the shape of a strategy of the initial strategy sets are considered. In particular, this prevents strategies from completely dying out. It is not desired to disturb the process too much by the emergence of mutants, such that the probability of emergence should be very small. We shall choose $\epsilon=10^{-7}$ or refrain from this disturbance and choose $\epsilon=0$.

BIMATRIX GAMES

The game between the four populations can also be seen as a *bimatrix game*. The following four 31×31 matrices describe four bimatrices:

$$A_{k,m} = (a(i,j;p_k,p_m), a(j,i;p_m,p_k))_{i,j\in N} \ \text{for} \ k=1,2, \ \text{and} \ m=3,4.$$

The first component in cell (i,j) of a bimatrix $A_{k,m}$ contains the payoff of strategy i of population p_k in a play with strategy j of population p_m, the second component of the tuple contains the payoff of the strategy j. Each of these bimatrices can be interpreted as a two-person game in normal form where one player chooses the row and the other player chooses the column. In our framework of the two-person game with incomplete information, we have to consider the following game. Consider a game between a row player of population p_k, k=1,2 and a column player of population p_m, m=3,4. The row player knows that with equal probability either matrix $A_{k,3}$ or matrix $A_{k,4}$ is chosen and the column player knows that with equal probability either matrix $A_{1,m}$ or matrix $A_{2,m}$ is chosen. We shall call this game the "*bimatrix game*".

A theoretical framework for the analysis of this bimatrix game is the evolutionary approach. The idea is that the evolutionary process optimizes fitness. The pioneer work in this field is done by Maynard Smith and Price (1973). Further developments in evolutionary stable strategies in bimatrix games and the introduction of this concept to extensive form games have been made by Selten (1983 and 1988). A concise mathematical discussion of the problem is given in Hofbauer and Sigmund (1988) and van Damme (1988). "An *evolutionary stable strategy* is a stable state of the process of evolution: if all individuals of a population adopt this strategy, then no mutant can invade" (van Damme, 1988, p.209).

Formally, our game has the structure of an "asymmetric animal conflict" (Selten 1980). Each player knows that he never plays against his own type (a first mover always plays a second mover and vice versa). Therefore the game has information asymmetry in the sense of Selten (1980). In such cases an evolutionarily stable strategy must be a pure strategy (Selten 1980).

14.2   RESULTS OF THE EVOLUTIONARY TOURNAMENT

For each of the three parameter values of α two evolutionary tournaments with the four populations were conducted. The two tournaments differed in the choice of the parameter ϵ: $\epsilon = 10^{-7}$ and $\epsilon = 0$. Each evolutionary tournament was stopped after 50,000 iterations. We shall report on the results of these tournaments in this section.

EVOLUTIONARY TOURNAMENT FOR $\alpha = 30$

Firstly we shall report on the evolutionary tournament with $\epsilon = 0$. The tournament yields stable weights after less than 600 iterations. These weights coincide for the two populations of first movers (p_1 and p_2) and the two populations of second movers (p_3 and p_4). The analysis of the average payoffs in the final tournament in Section 12.4 showed that the average payoffs of the two types of the first mover coincide in all but one case and for the two types of the second mover they coincide in all but two cases.

In populations p_1 and p_2 strategy 11 and strategy 14 die out, while the remaining strategies stabilize at a weight of .0345. Figure 14.1 shows the evolution of the weights of strategies 1, 11, and 14 for $t = 0,..,600$. Strategy 1 stands as a representative for the remaining strategies. The strategies 11 and 14 are exactly those strategies which do not demand 50 in the first decision step of the first mover. This high initial demand is only accepted by strategy 25. Strategy 11 has the highest average payoff in a single round tournament (see Section 12.4). In population p_3 and p_4 strategy 25 stabilizes on a weight of .006, and the other strategies have nearly equal weights at .0338. Figure 14.2 shows the evolution of the weights for strategy 25 and strategy 1 (as a representative of the others) in the first 600 iterations.

Strategy 25 is exploited by strategies 11 and 14, and on the other hand the first rank position of strategy 11 in a single round tournament is due to the low acceptance level of strategy 25. It is this interdependence which decreases the fitness of strategies 11, 14 (in p_1 and p_2), and 25 (in p_3 and p_4). However, after the enemies of strategies 25 died out, it is not able to increase its weight since it is not superior to the other strategies.

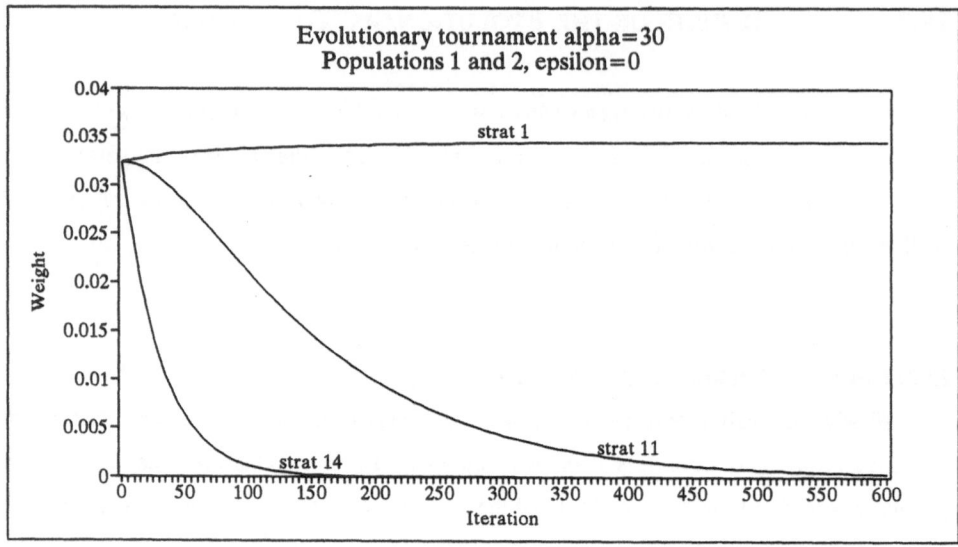

Figure 14.1:   Evolutionary tournament for $\alpha = 30$ and $\epsilon = 0$, populations p_1 and p_2

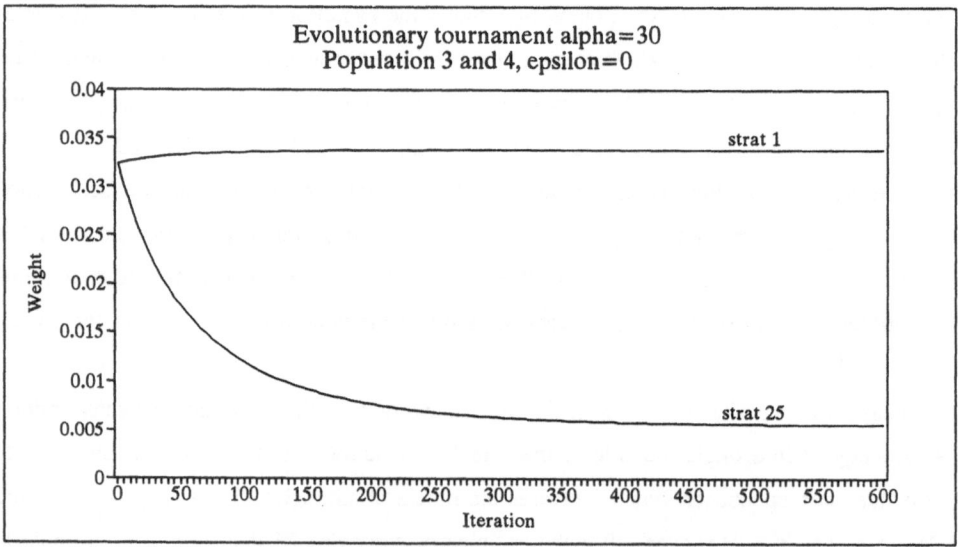

Figure 14.2:   Evolutionary tournament for $\alpha = 30$ and $\epsilon = 0$, populations p_3 and p_4

In an evolutionary tournament with $\epsilon=10^{-7}$ a very small number of mutants of each strategy emerges in each iteration. This prevents strategies from completely dying out or allows strategies to "recover" after the enemies fall back to this minimal weight. Such a recovering can be observed for strategy 25 in the populations 3 and 4. In the tournament with $\epsilon=0$ strategy 25 stabilized on a low weight after the enemies 11 and 14 died out. In a tournament with $\epsilon=10^{-7}$ the fitness of the strategies 11 and 14 deceases very soon to a value close to ϵ, and from this moment on strategy 25 is able to recover and gain weight. The remaining 28 strategies, which are represented by strategy 1 in figure 14.3 loose fitness in favor of strategy 25. Figure 14.3 shows the evolution of the weights over the 50,000 iterations. In population 2 and 3 there is no significant difference between the evolution processes of $\epsilon=0$ and $\epsilon=10^{-7}$.

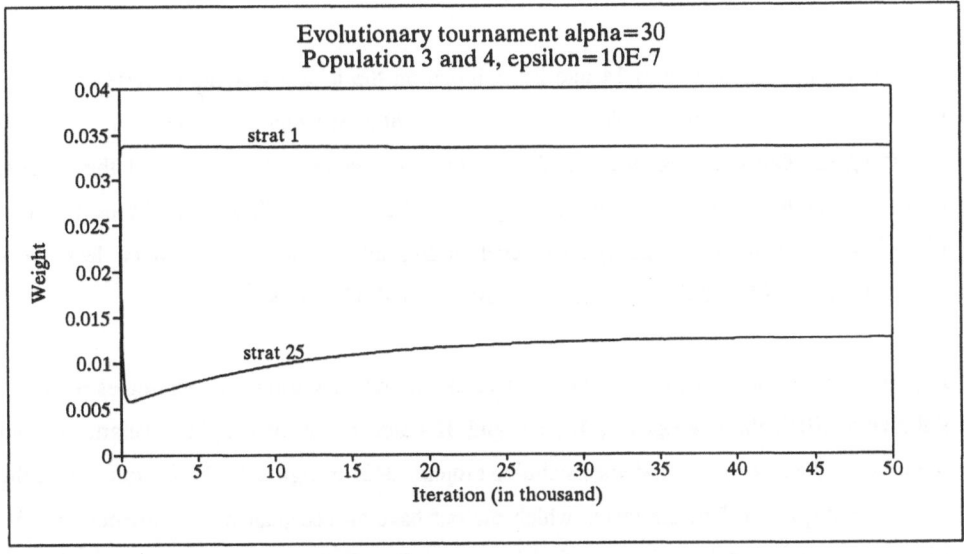

<u>Figure 14.3:</u>   Evolutionary tournament for $\alpha=30$ and $\epsilon=10^{-7}$, populations p_3 and p_4

EVOLUTIONARY TOURNAMENT FOR $\alpha=45$

The weights of the evolutionary tournament with $\epsilon=0$ stabilize after less than 300 periods. In population p_1 the strategies 1, 9, 11, 14, 17, 28, and 29 die out very soon. Strategy 25 reaches the highest weight with .0689. The strategies 5, 21, 26, 31, and 32 reach weights around .05, and the remaining strategies stabilize at a weight of .0344. Figure 14.4 shows the development of the weights for one representative of each of the above groups. The strategies which die out very soon are exactly those strategies which demand more than 50 in the first step of the first mover.

In population p_2 strategies 1, 9, 11, 14, 17, 26, 27, 28, 29, and 32 die out very soon, while the other strategies stabilize at a weight around .046. Figure 14.5 shows the evolution of the weights for one representative of each of the two groups. The strategies which die out are exactly those strategies which demand more than 50 in the first round of the first mover.

In population p_3 the strategies 14 and 25 stabilize on the lowest weights of .008 and .006, respectively. The strategies 5, 27, and 32 reach the highest weights around .054, while the remaining strategies stabilize around .034. Figure 14.6 shows the evolution of the weights for one representative of each of the three groups. The strategies 5, 27, and 32 are the ones which start with an initial acceptance level of 50, but lower the acceptance level very drastically in step 5 and 6, respectively (to 10, 30, and 35, respectively).

In population p_4 the strategies 26, 27, 28, and 29 die out very soon. The strategies 8 and 25 stabilize at .016, the strategies 7, 14, 17, and 32 reach the second highest weights around .033, while the remaining strategies stabilize around .042. In figure 14.7 the evolution of the weights is displayed. The strategies which die out have an acceptance level greater than 50 for the second mover in his first decision step. The further acceptance levels of these strategies are either constant or increase by time. The only other strategy with an acceptance level greater than 50 is strategy 1 with an acceptance level of constantly 51. Strategy 25 has an acceptance level of 46 in each step, and strategy 8 breaks off as soon as the opponent offers less than 50.

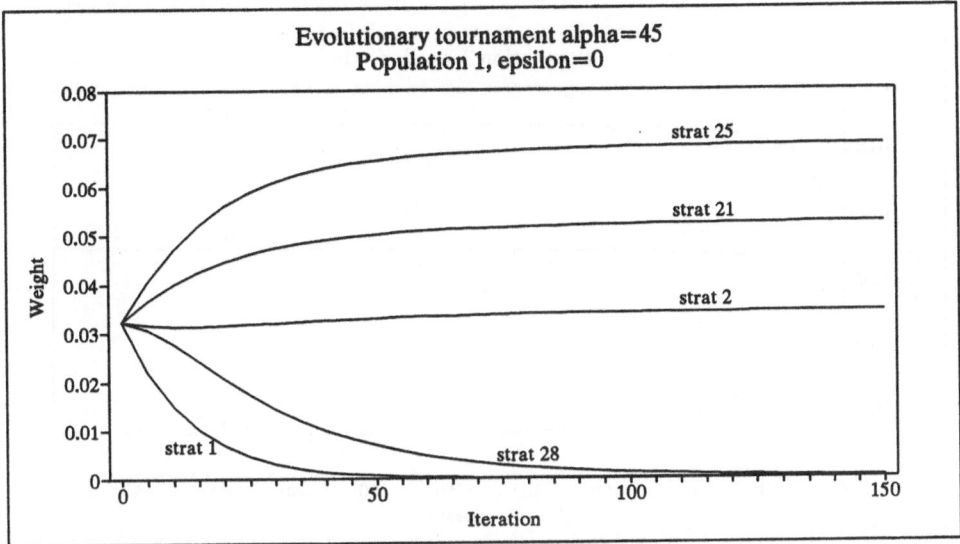

<u>Figure 14.4:</u>   Evolutionary tournament for $\alpha=45$ and $\epsilon=0$, population p_1

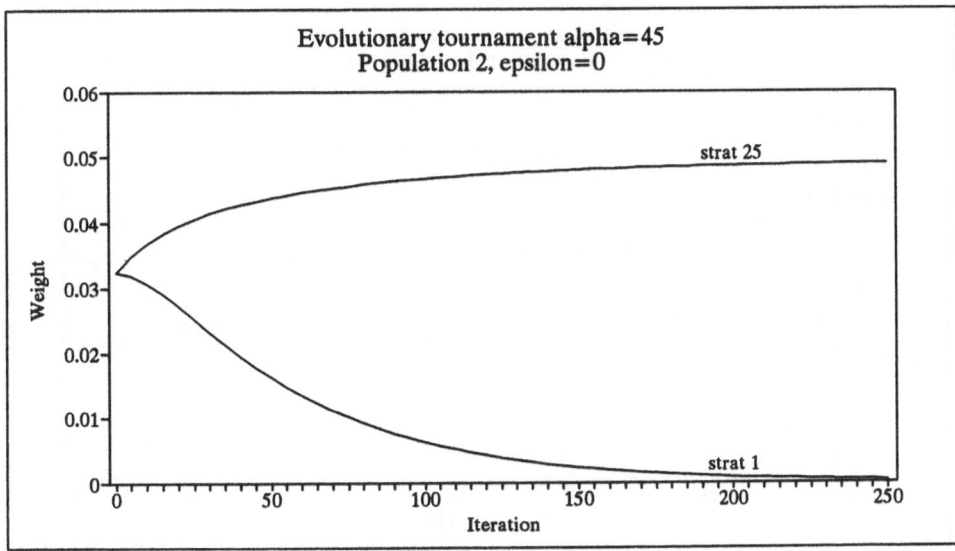

<u>Figure 14.5:</u>   Evolutionary tournament for $\alpha=45$ and $\epsilon=0$, population p_2

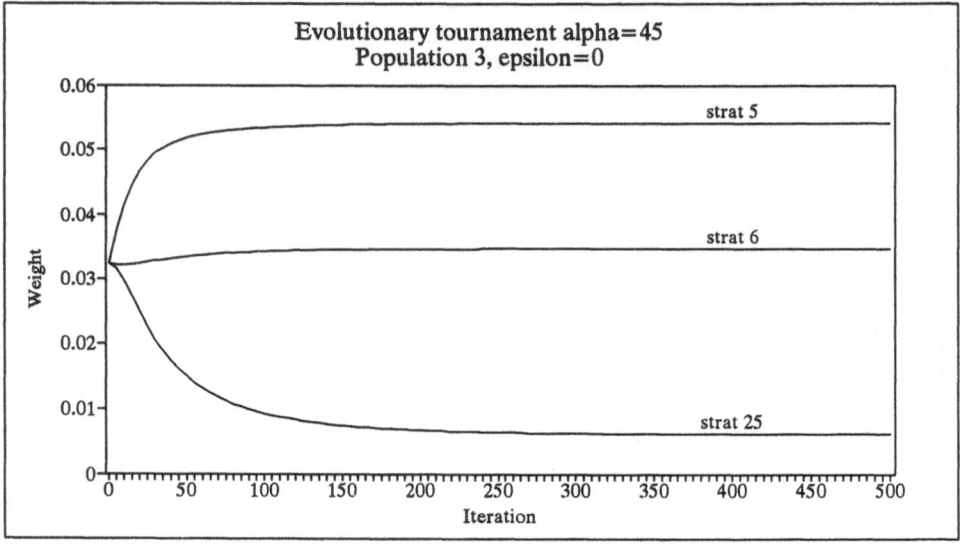

Figure 14.6: Evolutionary tournament for $\alpha=45$ and $\epsilon=0$, population p_3

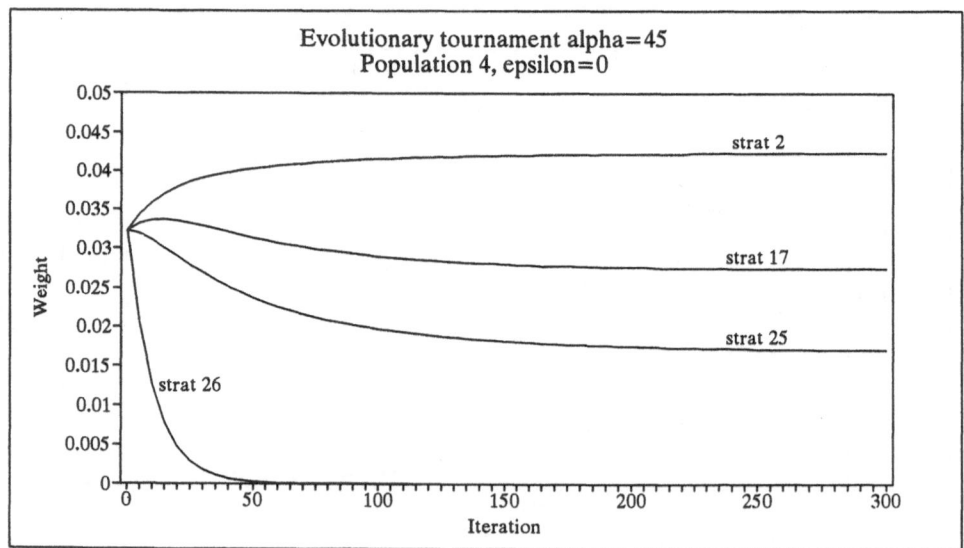

Figure 14.7: Evolutionary tournament for $\alpha=45$ and $\epsilon=0$, population p_4

The strategies with an initial demand greater than 50 die out in both populations of first movers. These strategies exploit the strategy 25 in both second mover situations. By the exploitation the fitness of strategy 25 decreases, and remains constant after the "enemies" died out. With the decrease of the weight of strategy 25 the exploiting strategies cannot survive anymore and die out. As a first mover strategy 25 can improve its initial weight. The strategy demands 50 in each decision step, which is mostly accepted by the opponent. If not, strategy 25 accepts in the third step every proposal which yields an individually rational outcome. As a first mover the strategy is not exploited by high initial demands of the opponent, but in case no immediate agreement is reached, it mostly agrees in the third decision step, which is more profitable than having a conflict or a long bargaining.

The strategy for the high type of strategy 17 demands 54 in the first six decision steps, which is accepted by strategy 25. The high weight of strategy 25 as a first mover yields an only moderate decrease in the weight of strategy 17.

For the low type second mover the strategies with a drastic decrease in the acceptance level reach the highest fitness. Initially, they agree to every proposal yielding more than 50, but in step 5 or 6 they decrease to an acceptance level of 10, 30, or 35. These are the sharpest drops in acceptance levels, but this seems to be successful in order to avoid a conflict in a play with a high type. For the low type first mover the strategies 5, 21, 26, 31, and 32 have a slight advantage by a sharp drop in the acceptance level.

If one looks at the evolutionary tournament with $\epsilon = 10^{-7}$ the main difference to the tournament which does not allow the emergence of mutants occurs in the evolution of strategy 25 (like for $\alpha = 30$). The strategies which demand more than 50 in the first decision step exploit this strategy and therefore its weight decreases. However, strategy 25 is able to recover after the enemies fall back to a weight close to ϵ. The emergence of mutants then allows an increase of the fitness of strategy 25. This is observable in the populations 1, 3, and 4, as figure 14.8 shows.

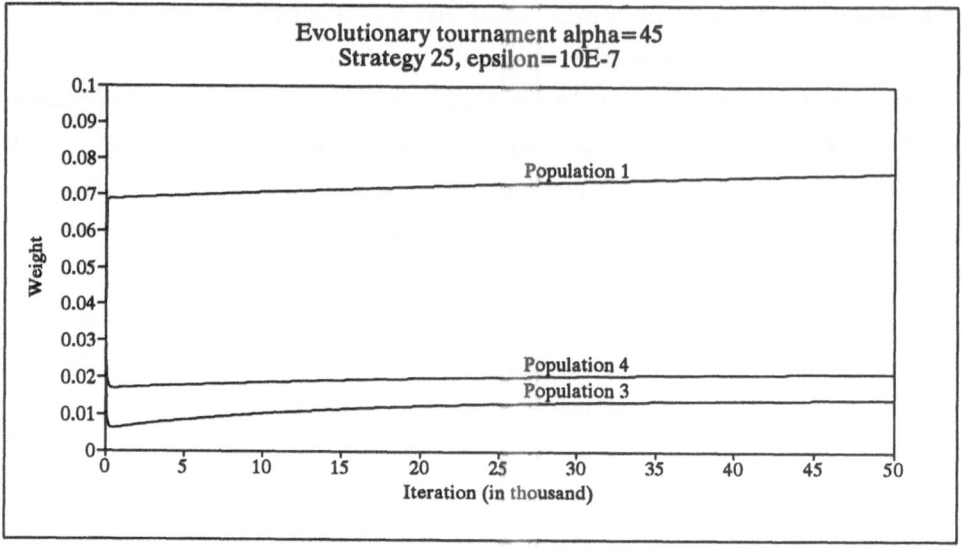

<u>Figure 14.8:</u>   Evolutionary tournament for $\alpha=45$ and $\epsilon=10^{-7}$, strategy 25

EVOLUTIONARY TOURNAMENT FOR $\alpha=60$

The stabilization of the results of the evolutionary tournament with $\epsilon=0$ takes considerably longer than in the cases of $\alpha=30$ and $\alpha=45$. It can be observed after roughly 4,000
iterations. Another difference to the previous tournaments is that the weight processes of the
strategies are no longer monotonic for $\alpha=60$. It is also observable that the selection process
is more strict. After the stabilization of the evolutionary process only between 1 and 3
strategies survive. The evolution processes conducted with a small ϵ of 10^{-7} in the replicator
dynamics look considerably different. Here, the small probability of an emergence of a
mutant does not stabilize the processes of the low types, but leads to cyclic movements.

At first we shall discuss the results of the evolutionary tournament with $\epsilon=0$.

In population p_i only two strategies survive: strategy 2 with a weight of .0071 and strategy
4 with a weight of .9930. Strategy 4 plays very tough as a low type first mover. The
succession of the demands is: 80, 75, 70, and 50 from the 7th step on. Also the acceptance
level is very high: 75, 70, and 50 from the 6th step on. Also strategy 2 plays tough with a

demand and an acceptance level of 80 in the first three steps. Besides these two strategies there is only one other strategy with an initial demand and an initial acceptance level of 80. But this strategy decreases its demands and its acceptance level in an opponent dependent way. Therefore, one can say that the strategies 2 and 4 play the toughest initial phase of the bargaining.

In population p_2 the three strategies 7 (with a weight of .4122), 17 (with a weight of .2145), and 31 (with a weight of .3733) survive. Contrary to all other strategies which die out very soon, the strategies 3 and 30 die out in iteration 4,000. All three surviving strategies demand 80 in the first decision step of the first mover. Strategies 7 and 31 decrease their demands in an only step dependent order to 70, and strategy 17 decreases the demand in an opponent dependent way. Strategy 7 has a constant acceptance level of 61, strategy 31 has a constant acceptance level of 65, and strategy 17 starts with an acceptance level of 65 and decreases it opponent dependently. The break off prescription of strategy 31 is only step dependent, while the break off prescriptions of the strategies 7 and 17 depend on the last proposal of the opponent.

In population p_3 only strategy 28 survives, after it "won a big fight" with strategy 17. This fight drives the fitness of these two strategies up and down in a cyclic way. All other strategies die out soon. The strategies 17 and 28 are exactly those strategies which make the first demand as a second mover dependent on the initial demand of the first mover. In case the opponent proposed 20 or less strategy 17 plays very weak: it has an acceptance level of 30 and proposes 29. Strategy 28 conditions its actions on the proposals of the opponent. If the opponent makes only small proposals the strategy plays weak with low demands and a low acceptance level. These two strategies are very flexible and play soft in case the opponent plays tough.

In population p_4 three strategies survive: strategy 26 at a weight of .71793, strategy 29 at a weight of .27725, and strategy 32 with the weight of .00480. The other strategies either die out very soon, or close to iteration 300. Strategy 26 starts with an initial demand of 80 and decreases it in a step dependent order to 75. It has a constant acceptance level of 75 and a break off prescription which depends on the last proposal of the opponent with an upper step limit. Strategy 29 specifies an initial demand of 75 and decreases it in an opponent dependent order. It also has a constant acceptance level of 75 and the break off prescription conditions on the last proposal of the opponent.

The evolution of the weights of the four populations for $\epsilon = 0$ are shown in figures 14.9 to 14.12.

With a decrease in the weight of strategy 4 in population p_1 an increase in the weight of strategy 17 in population p_3 occurs. With strategy 4 also strategy 2 of population p_1 and strategy 28 of population p_3 decrease. The tough low type first movers, strategies 2 and 4, exploit the very soft strategy 17. Finally, the also flexible but less soft strategy 28 (with an acceptance level of 50, in comparison to 30 for strategy 17) survives as a low type second mover.

If one allows the emergence of mutants with a small probability of 10^{-7} the results look different, as shown in figures 14.13 to 14.16. But, except for population p_1 the same strategies are of importance. In population p_1 strategy 28 alternates cyclically with strategy 4, while strategy 2 falls back to a very small constant weight. For both low types the evolutionary process does not stabilize after 50,000 iterations. In population p_2 the same strategies remain important, but with different weights. In population p_4 only strategy 26 survives.

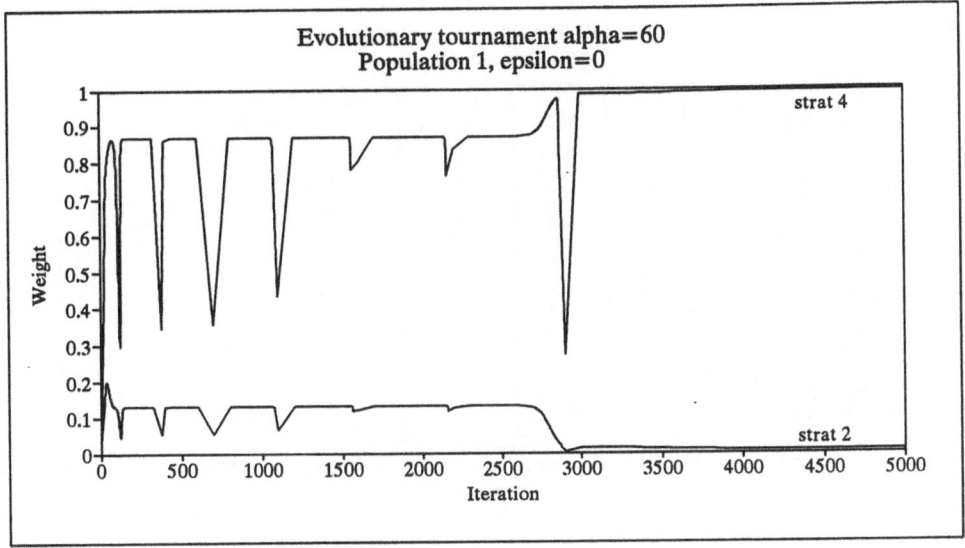

Figure 14.9:   Evolutionary tournament for $\alpha=60$ and $\epsilon=0$, population p_1

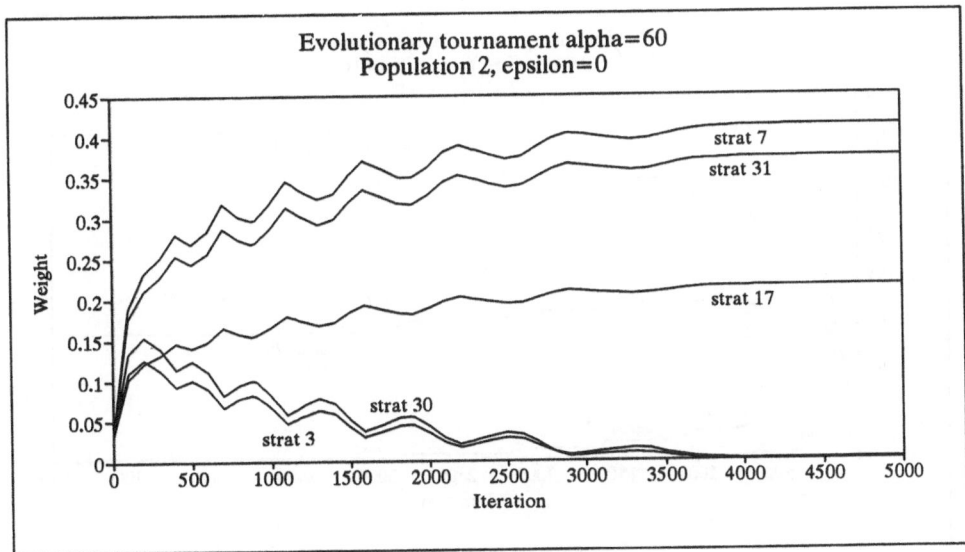

Figure 14.10:   Evolutionary tournament for $\alpha=60$ and $\epsilon=0$, population p_2

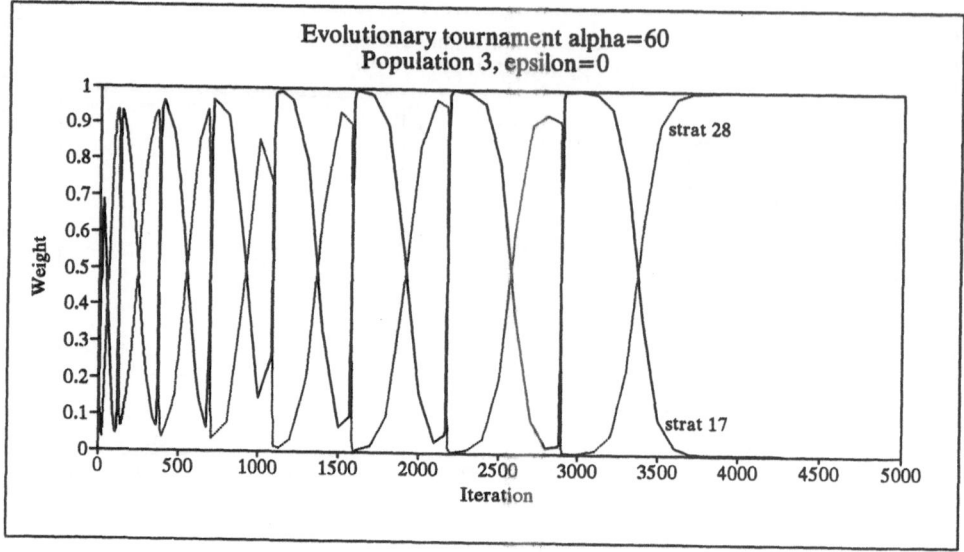

Figure 14.11:   Evolutionary tournament for $\alpha = 60$ and $\epsilon = 0$, population p_3

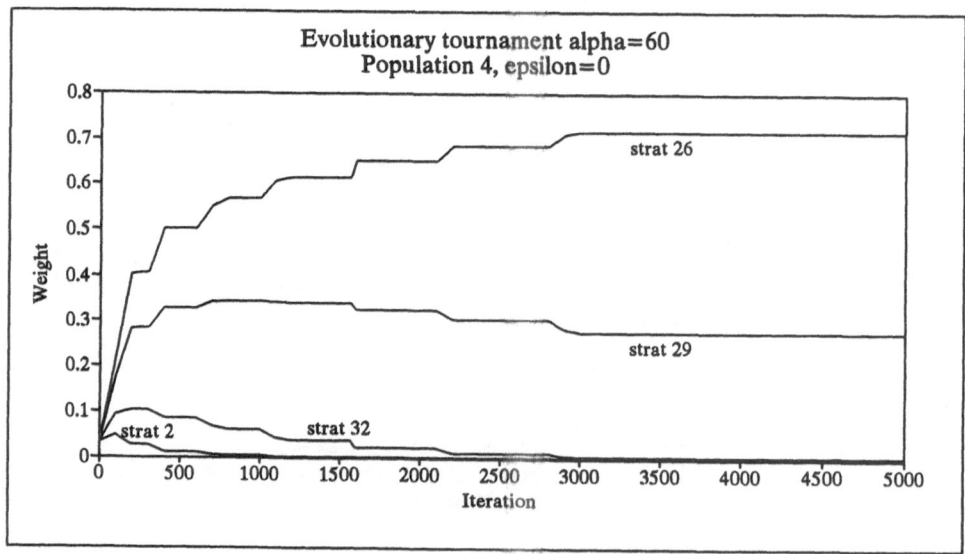

Figure 14.12:   Evolutionary tournament for $\alpha = 60$ and $\epsilon = 0$, population p_4

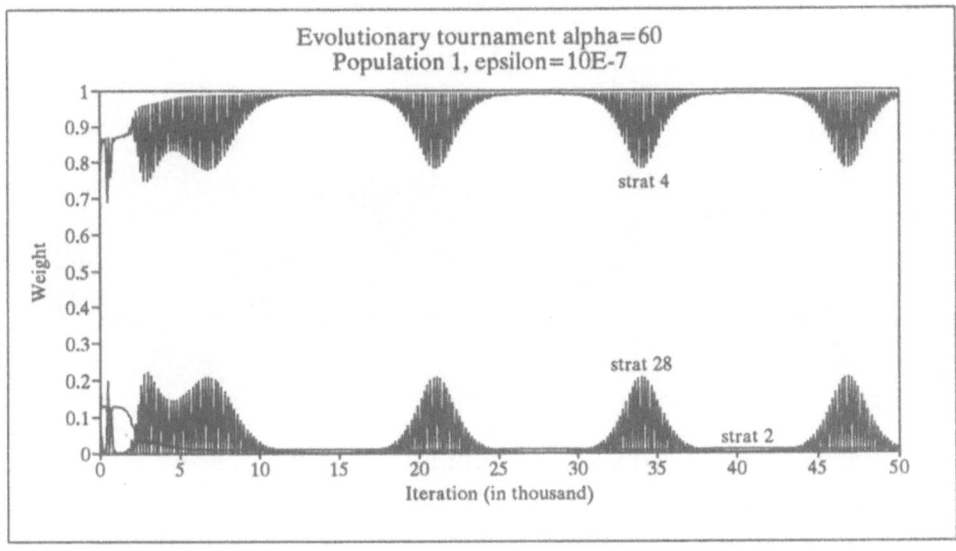

Figure 14.13:   Evolutionary tournament for $\alpha=60$ and $\epsilon=10^{-7}$, population p_1

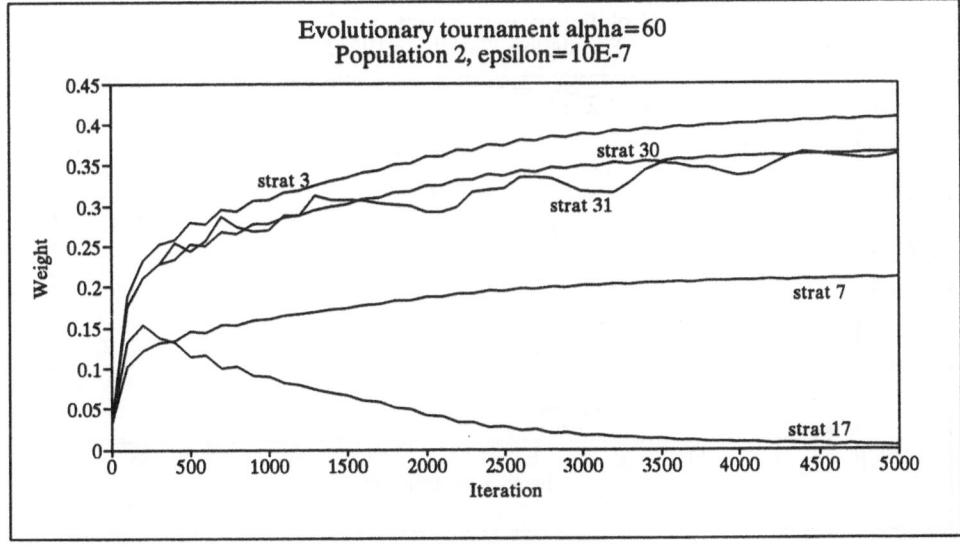

Figure 14.14:   Evolutionary tournament for $\alpha=60$ and $\epsilon=10^{-7}$, population p_2

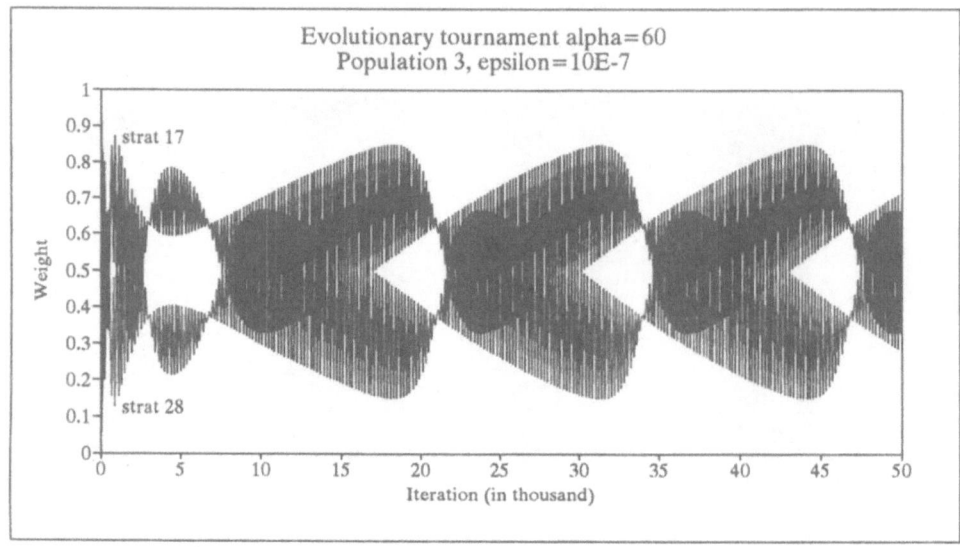

Figure 14.15:   Evolutionary tournament for $\alpha=60$ and $\epsilon=10^{-7}$, population p_3

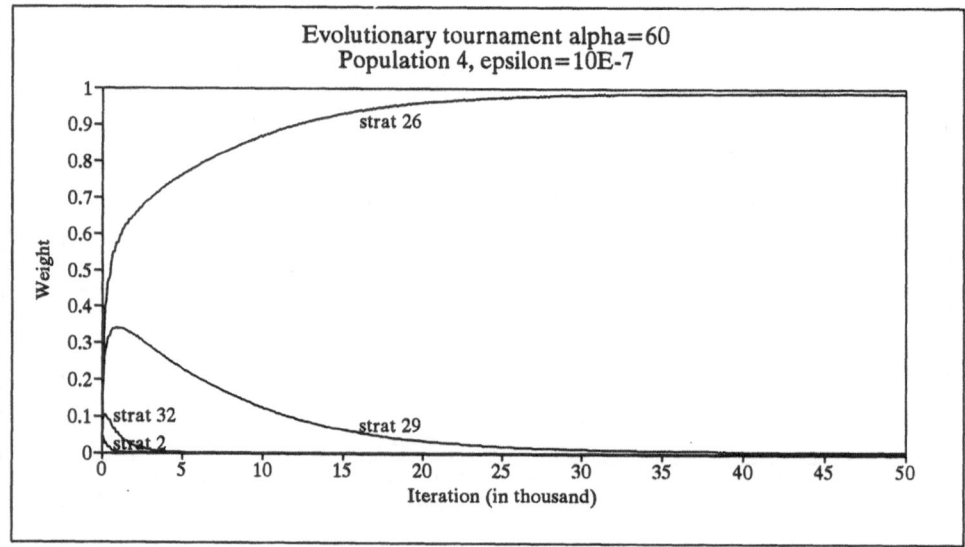

Figure 14.16:   Evolutionary tournament for $\alpha=60$ and $\epsilon=10^{-7}$, population p_4

TIME AVERAGES

Even if an evolutionary process like the one studied here does not converge, the time average of the process may converge to an equilibrium point in mixed strategies. It has been shown in the literature that under certain conditions this is the case for the continuous replicator process applied to two-person games in normal form (Schuster, Sigmund, Hofbauer, and Wolff 1981, and Hofbauer and Sigmund 1988). Therefore, it is of interest to explore the question whether the same convergence of time averages is observed in the case at hand.

For each α we shall examine whether the average weights over the 50,000 iterations are equilibria of the bimatrix game defined in Section 14.1. Unfortunately, the problem of singularity of the system of linear equations, which has to be solved for this purpose, emerges. This problem is caused by strategies of different populations which reach the same average payoff in plays with another population. An example is that a low type first mover reaches the same payoffs in plays with the (remaining) strategies of the second mover populations than a high type first mover. This causes two identical rows in the matrix of the linear equation system. This problem cannot be solved by removing one of the rows since the underlying strategies stem from different populations. In particular this means that there is a variety of equilibria.

Nevertheless, it is possible to examine whether the time average weights of the evolutionary tournaments are in equilibrium, at least approximately. This means that the system of linear equations for the determination of the equilibrium weights is evaluated with these average weights in order to obtain a candidate for an "equilibrium" payoff for each strategy. This evaluation shows that all strategies which survive in the evolution process have the highest payoffs and that these payoffs are equal. Only for $\alpha=60$ in two populations differences in the second decimal digit of the payoffs (which are numbers between 0 and 100) occur. This means that the average weights over the 50,000 iterations solve the system of linear equations and therefore form a mixed strategy equilibrium (at least approximately).

14.3   FITNESS AND TYPICITY

Is there a relationship between the typical strategies and the strategies which survive in the evolutionary tournament? Is there even a correlation between typicity and fitness of a strategy? In this section we shall investigate these questions.

For $\alpha=30$ there are only three strategies distinguished from the remaining ones by their fitness. These are the strategies 11 and 14 of the two first mover populations, and strategy 25 of the two second mover populations. Strategies 11 and 14 die out, while strategy 25 stabilizes on a very low weight. These are also strategies with a very low typicity. Therefore, it is not surprising that a Spearman rank correlation analysis between the typicity of a strategy and its average fitness over the 50,000 iterations yields a positive correlation for all four populations (three times at a significance level of .01, and once at .1, one-sided). The main misconception of strategies 11 and 14 is that they are not able to reach an immediate equal split agreement in case they are first mover. Therefore, their survival depends on strategies they are able to exploit. If they do not exist, these strategies die out. It is as well typical as evolutionary stable to be able to reach immediate equal split agreements.

For $\alpha=45$ a very similar picture emerges. Also here the strategies which are not able to reach immediate equal split agreements as first movers die out in the evolutionary process. As a low type second mover the strategies which initially strive for the equal split agreement, but if this is not reachable, play very soft, reach the highest fitness. The Spearman rank correlation analysis between the typicity of a strategy and its average fitness can only find a correlation for the two high types. Here, in both cases a positive correlation is found at a significance level of .01 and .05, respectively. For the two low type players no significant correlation can be found. However, it is true that the characteristic concerning the acceptance level of the low type player specifies a constant or a decreasing acceptance level. Since in most of the strategies this characteristic is present in the form of a constant acceptance level, the correlation is to weak to be significant.

In the evolutionary tournament with $\alpha=60$ the two low type first mover strategies with the toughest initial phase survive. They are typical with respect to the demand behavior and the non-existence of a break off prescription. The acceptance prescription, however, is not typical. It is considerably higher than observed in the majority of the strategies. For the low type second mover the two most flexible strategies survive the evolution. It is not typical to specify the demand in an opponent dependent way, therefore both strategies do not have characteristic 2. It is also not typical to decrease the acceptance level in an opponent dependent way. Therefore, in this respect the strategies are not typical. The Spearman rank correlation analysis cannot find a correlation between the typicity and the average fitness over the 50,000 iterations (with $\epsilon = 0$) for population p_1 as well as for population p_3 at a significance level of .2 (two-sided). Accordingly, tough imitation strategies survive for the low type first mover and flexible opponent dependent strategies survive for the low type second mover.

For the high type first mover of $\alpha = 60$ three strategies with high, medium and low typicity survive. The Spearman rank correlation cannot find a correlation between the typicity and the average fitness. For the high type second mover, the surviving strategies have medium or low typicities, but the Spearman rank correlation does not detect any correlation between the typicity and the average fitness.

The participants of the strategy experiment did not put much emphasis on the distinction between the first and the second mover situation. The distinction emerged implicitly since the first mover has to act in all odd steps and the second mover acts in all even steps. For $\alpha = 30$ a typical strategy only specifies a step dependent (and therefore first and second mover distinguishing) break off prescription for the high type. For $\alpha = 45$ only the acceptance level of a typical strategy for the low type and the break off prescription in a typical strategy of a high type are possibly step dependent. In the evolutionary tournament, however, the roles of the first and the second mover polarize. One becomes the exploiting and the other one the exploited type. We observe that the first mover strategy and the second mover strategy of a participant have different average weights. Indeed, it may happen that one of them dies out while the other one survives. The repeated play of the game in the evolutionary tournament makes the implicitly specified differences of the mover situations visible.

CHAPTER 15.   SUMMARY AND CONCLUSIONS

Finally, we shall review the main results gained in this book. In particular we shall link the results of the game theoretical analysis and two different experiments, the game playing experiment and the strategy experiment. For a more detailed summary of the results of the game playing experiment see Chapter 11.

The *game playing experiment* explored the spontaneous behavior of subjects in the two-person bargaining game with incomplete information. The experiment was designed such that each subject played 16 plays and the experimenter could distinguish four levels of experience for each gametype. This allows to study the learning of the subjects from play to play, and to distinguish between inexperienced and experienced behavior of subjects in the same game situation. In contrast, the aim of the *strategy experiment* was to explore the strategies programmed by highly experienced subjects. We know that a strategy does not reflect the full diversity of a subject's behavior. The developer of a strategy has to concentrate on the (from his point of view) most relevant features of the problem. Therefore, strategies yield a more tight picture of the behavior. Since the strategies were not allowed to have a memory, they could not learn from the previous plays.

DEMAND AND ACCEPTANCE BEHAVIOR

All strategies in the strategy experiment specify the acceptance decision in the form of an *acceptance level*. They specify a bound and accept every proposal above this bound.

For $\alpha < 50$ a strategy of the strategy experiment typically does not try to identify the type of the opponent. The opponent's last proposal is evaluated for the acceptance decision, but the demand behavior does not depend on the opponent's proposals. Typically a demand of 50 is specified for each step. For $\alpha = 30$ we typically observe an acceptance level of 50, which remains unchanged for all steps. For $\alpha = 45$ the initial acceptance level typically is 50 and it remains on this level or decreases in a step dependent way.

For $\alpha = 60$, in contrast, we more frequently observe strategies which classify the opponent. The demands as well as the acceptance levels depend on this classification. However, an opponent dependent adaptation does not occur in the relevant typical characteristic. For the low type as well as for the high type the initial demand typically is 70 or 80, which is in the same range as in the game playing experiment. The further demands of the low type typically decrease in a step dependent way and the further demands of the high type typically are identical to the initial one. The initial acceptance level of the low type typically is 50 and the further acceptance levels are either identical to the initial one or decrease in an only step dependent way. For the high type the typical initial acceptance level is between 60 and 70 and the further acceptance levels are identical to the initial one.

In the game playing experiment we found more than 70% of the demands on prominent numbers (numbers divisible by 5 without a remainder). In the strategy seminar we also observe a preference for prominent numbers in the demands. The prominence of 50 causes the high number of prominent demands for $\alpha < 50$. For $\alpha = 60$ the initial demand typically is prominent. The further demands of the strategies are mostly on prominent numbers.

For $\alpha < 50$ a strategy's demand and acceptance level in a particular step typically coincide. For $\alpha = 60$ we observe that the demand of a bargaining step is considerably higher than the acceptance level. The strategy tries to reach a high agreement outcome by tough demands, but on the other hand does not want to miss the chance of a "reasonable" agreement.

The strategies which specify demands according to a predetermined list, specify in all but one case a decreasing list. Typically the decrease is not regular in the step number. A strategy specifies the same demand for several steps, then decreases the demand considerably and repeats it for several steps. The same observation can be made for the acceptance behavior. No strategy specifies threats (sharp demand increases), although this is frequently observed in game playing experiments. An explanation for this phenomenon is that an opponent's strategy would not notice a threat. The opponent's acceptance level ensures that a threat is not accepted and the opponent's demand is either not contingent on the player's last proposal, or it depends on a classification of the player's proposal in certain ranges. But these ranges are not so tight that they could distinguish between high demands and threats. Actually, the participants of the strategy experiment were ignorant about the strategies of the other participants, but either they concluded from their own strategy or from experience that threats are not worthwhile.

THE AGREEMENT OUTCOMES

A general observation is that the strong player receives a lower agreement outcome than in a comparable game under complete information (Kuon and Uhlich, 1993). During the bargaining process the strong player has no possibility to prove his strength and on the other hand there is no instrument to detect whether the opponent is weak. Therefore, a tough bargaining behavior might always be interpreted as a "bluffing" weak player. Hence, the incomplete information increases the bargaining power of the weak player and decreases the bargaining power of the strong player (in comparison to a game with complete information).

The experienced subjects of the game playing experiment agree on the equal split for $\alpha < 50$. Exactly the same observation can be made in the strategy experiment. In both experiments the agreements occur very quickly. Although game theory also prescribes other Nash equilibrium outcomes for the case of $\alpha < 50$, the two kinds of experiments clearly single out the equal split as the allocation scheme. The equal split is a focal point which is individually rational for both types of players and due to the incomplete information the strong player seems to have not enough bargaining power to achieve a higher outcome.

In the game playing experiment the agreement outcomes for games with $\alpha > 50$ become more asymmetric with a higher level of experience. This is especially true for plays of two weak types. Typically one weak type successfully imitates a strong type such that the agreement outcomes of two weak types look like the agreement outcomes of a weak and a strong type. The most successful of the tested prediction concepts for the agreement outcomes of the low type predicts all prominent numbers above the to the adjacent prominent number adjusted equal split of the difference. In the strategy experiment only seven of the 31 strategies demand the equal split as a first mover. The other strategies "bluff" by initial demands of typically 70 or 80. Although no weak opponent's strategy accepts such a high demand, highly asymmetric agreements also occur. They are caused by a strategy which lowers its acceptance level more quickly or more drastically than the opponent's strategy. Typically the asymmetric outcomes are not reached by low demands of one strategy.

Also in a game with $\alpha > 50$ the strong player's agreement outcome is lower than in a comparable game with complete information. In the game playing experiment the most successful of the tested prediction concepts for the agreement outcomes was the *prominent fair distribution scheme*. It specifies $\alpha+5$ as the strong player's lower outcome bound. In the strategy experiment the initial acceptance level of a strong player's strategy is typically between 60 and 70, and this acceptance level remains identical or furthermore decreases. The demands of the strong player's strategy typically decrease to 70. This confirms that the strong type is satisfied with only a small surplus in addition to his alternative value.

In the game playing experiment for games with $\alpha > 50$ we detected in about one third of all plays a phenomenon called *sudden acceptance*. It describes that from a high demand level of both players, one player suddenly accepts the proposal of the opponent. An acceptance is defined as sudden acceptance if the acceptor receives at least 10 less than his lowest demand. This phenomenon is also found in the strategy experiment. It was already mentioned that the acceptance level is considerably lower than the demand level in case of $\alpha=60$, and that the decrease in the acceptance level occurs by significant drops after constant periods. Therefore, it happens that on a high demand level all of a sudden an agreement may occur.

THE AVERAGE PAYOFF

For $\alpha > 50$ the strong player does, on average, not achieve an individually rational payoff in the game playing experiment. This means that the high type first mover receives a payoff lower than α and the high type second mover receives a payoff lower than $\delta\alpha$. This is the payoff a strong player can guarantee himself by a unilateral break off in his first decision step. A non-individually rational payoff contradicts the assumption that the players play a Nash equilibrium. This observation can be made in the low as well as in the high experience levels and does not change significantly with experience. The gain in addition to the strong player's alternative in case of agreement is too low to compensate the losses he makes from break offs after a considerable number of steps.

In the strategy experiment we find ten strategies which do not achieve individually rational payoffs as a first mover and only one strategy which does not achieve an individually rational payoff as a second mover. With the different tournaments the strategies became more moderate in their bargaining behavior and were able to achieve higher average payoffs.

If a strong player had noticed that he is not able to achieve an individually rational payoff, he consequently could break off in his first decision step. In the game playing experiment, however, the number of immediate break offs does not rise with experience. Interestingly, four strategies in the strategy seminar break off immediately as a strong type.

BREAK OFFS

In the strategy experiment a low type strategy typically does not specify a break off condition. This, however, does not mean that a low type strategy always agrees. The participants of the strategy experiment knew that each bargaining game which was not terminated by break off or acceptance up to a certain step was terminated by the tournament program with zero payoffs for both strategies. This step was determined as a randomly chosen step after the discounted coalition value was below .01. This means it was a step where the coalition value was below the computation precision. Therefore, a strategy was not forced to determine a break off condition. For a strategy there is no "weariness of bargaining", therefore we do not observe "active" break offs by the weak types, like they can be observed in the game playing experiment. Nevertheless, non-agreements in the form of "time expiration" occurred.

In the game playing experiment the major reason for a break off by the strong player in a game with $\alpha < 50$ was that the opponent did not offer at least 50 during the bargaining. The strategies of the strategy experiment reflect the same principle. The break off of the strong player in games with $\alpha < 50$ typically occurs after a predetermined number of steps. If the strategy was not able to agree before this step, it breaks off. Since a strategy's acceptance level typically is 50 we observe the same behavior as in the game playing experiment.

The main reason for the break off of the strong player in the game playing experiment with $\alpha > 50$ was that the strong player did not receive an individually rational proposal. In the strategies of the strategy experiment we typically find a break off if the opponent's last proposal is lower than a predetermined bound, which mostly is 60. This means that the strategy breaks off as soon as the opponent does not make an individually rational proposal. This is a much tougher criterion than used in the game playing experiment. The strategies do not wait for the opponent to increase his proposal. This explains why the strategies achieve higher average payoffs for the strong type than the subjects in the game playing experiment.

The subjects are too patient in hoping for an individually rational agreement, such that the additional gain a strong player receives from an agreement is too low to compensate the discount loss from a "too late" break off. A minority of 13 strategies specifies the break off only in dependence of the step number. This behavior is comparable to the one observed in the game playing experiment.

REVELATION OF THE WEAK ($\alpha > 50$)

In the game playing experiment it was found that a revelation of the weak player is profitable in a play with a strong player, in the sense that a revealing weak player receives on average a higher payoff than a non-revealing weak player. On the other hand, if the opponent is weak too, a revelation is not advantageous since the revealer than is "exploited". Overall, a revelation by the weak player does not pay for $\alpha=55$ and $\alpha=60$, but for $\alpha=70$ the revealer receives a higher average payoff. For $\alpha=60$ about twice as many non-revealers as revealers were observed.

In the strategy experiment we observe seven strategies which reveal by a demand of 50 in the first decision step of the first mover. All but four strategies reveal their type during the demand process, but most of the strategies do not make demands which leave individually rational amounts for a strong opponent. This is a phenomenon which was also observed in the game playing experiments. This mostly results in a break off, or by a sudden acceptance of the weak player an agreement is reached.

THE BARGAINING PROCESS

In the strategy experiment we seldom observe a bargaining process for $\alpha<50$. The typical strategy demands 50 in the first decision step and has an initial acceptance level of 50. The bargaining process which is implemented in the strategies typically is independent of the opponent's proposals. It specifies fixed goals for fixed steps and if it is not possible to achieve these goals with the opponent, the bargaining fails, either by break off or by non-agreement at the time of termination by the tournament program. No attempt is made to detect the opponent's goals in order to strive for a compromise. In the game playing experiment it was shown that the bargaining process of a subject can successfully be described by an exponential process. This means that the subject makes positive concessions which shrink over time, independent of opponent's proposals. This is similar to the behavior of the strategies in the case of $\alpha < 50$: fixed goals which the subjects want to achieve and either they are reachable with the opponent or the bargaining fails.

For $\alpha = 60$ we observe more strategies in the strategy experiment which condition on the proposals of the opponent, but most frequently a step dependent adaptation of the demand and the acceptance level is observed. As mentioned before, the acceptance level is held constant for several steps and if no agreement can be reached at a certain level it is adapted (lowered). The acceptance levels can be seen as aspiration levels of the strategies. In the game playing experiment we found points of stronger resistance in the concession process of the subjects. It was observed that the concessions of the players shrink overproportionally as certain bounds are approached. These bounds, the numbers divisible by 10 without a remainder and especially the value 50, were interpreted as natural choices of aspiration levels of the subjects. The subjects plan to achieve a ceratin goal, but as they see that it is not reachable, they abandon it and form a new goal.

LEARNING

The strategies of the strategy experiment were not allowed to have a memory such that they were not able to learn from previous plays. The only learning we could observe in the strategy experiment is the change of the strategies after the tournaments. The participants of the strategy seminar wrote in the final seminar paper that they weakened the bargaining behavior, which in particular means they lowered the demands and the acceptance levels from tournament to tournament since they observed a large number of "avoidable" conflicts.

In the game playing experiment we can observe the change in the demand behavior after the experience of the previous play. A tougher bargaining behavior can be found after an agreement in the previous play and a weaker bargaining behavior after a conflict. With tougher bargaining behavior it is meant that the initial demand increases and the concessions decrease. A weaker bargaining behavior is characterized by a lower initial demand and higher concessions in comparison to the previous play. After an agreement the player "tests" whether with a tougher bargaining also an agreement is reachable. If the previous play ended in conflict, the player weakens his bargaining in order to avoid a conflict in this play. For $\alpha < 50$ the change in the bargaining behavior after an agreement was less drastic than after a conflict. An explanation is that a conflict gives a much clearer signal of failure than an agreement gives hope for a better agreement. For $\alpha > 50$ the two changes have approximately the same magnitude. Here, a conflict is not necessarily a signal of failure.

THE WINNING STRATEGY

The strategy winning the final tournament of the strategy experiment prescribed the most moderate bargaining behavior observed in the experiment. For $\alpha < 50$ it always proposes 50 and has an acceptance level of the alternative value plus 1. This means it accepts every proposal yielding strictly more than the alternative value (which is 1 for the low type). For the low type of $\alpha = 60$ the strategy demands 50 for the first and the second mover. Except for step 2, it accepts every proposal greater or equal to 1 and breaks off only if the opponent proposes 0. For the high type of $\alpha = 60$ the strategy demands 80 as a first mover, accepts every proposal greater than 60 and breaks off at a proposal of lower or equal to 60.

This strategy reached the highest payoff in the final tournament since it was not really exploited by other strategies in the sense that they reach very high outcomes like 99. A small number of strategies was able to achieve agreement outcomes like 65 in a play of a game with $\alpha < 50$ against the winning strategy, but on the other hand this "loss by exploitation" was small enough to be compensated by the gain which resulted from the shortness of the bargainings ending in agreement and from the small number of break offs.

Such a weak acceptance behavior was not observed in another strategy and never observed in the game playing experiment. The developer of a strategy specifies certain bounds above the alternative value which should be reached by an agreement. If the bargaining is not able to satisfy these goals it is terminated by conflict, even if it is less profitable than accepting the last proposal by the opponent. This is remarkable since there is nothing like "satisfaction about the exploitation of the opponent" or "loss of face" in a strategy. The tournaments were completely anonymous and the participants of the strategy experiment knew that the payoff of the final tournament was the major component of the final grade.

THE SEQUENTIAL EQUILIBRIUM SELECTED BY CHATTERJEE AND SAMUELSON (1988)

In Section 2.2 we described the sequential equilibrium Chatterjee and Samuelson (1988) selected for an equivalent buyer-seller problem. Remember, that the pure strategy of the strong player prescribes a demand of the unique sequential equilibrium demand of the complete information game against a weak opponent (*concealing demand*) and accepting no offer worse than this. This means that the strong player does not loose bargaining strength by the introduction of incomplete information to the game. The game playing experiment as well as the strategy experiment do not confirm this assumption. In both experiments the strong player was satisfied with lower payoffs than in the complete information game.

For the weak player the sequential equilibrium prescribes a randomization between the concealing and a revealing demand. If the probability expressing the likelihood of a weak opponent reaches a critical value, the weak player either accepts the concealing offer or makes a revealing demand. The sudden acceptance can be seen as the acceptance of the concealing offer after the probability reached the critical value. The weak player all of a sudden accepts the opponent's offer after a series of high demands. The case of a revealing demand by the weak player initiates a quasi subgame of one-sided incomplete information. The uninformed player makes a series of increasingly favorable offers to the informed player which are chosen to make the weak informed player indifferent between accepting and waiting for the next more favorable offer. The weak informed player randomizes between accepting and rejecting these offers. The informed players make concealing offers which are rejected by the uninformed player. The uninformed player revises downwards the probability describing the likelihood that the opponent is weak. As this probability reaches a critical value the uninformed player accepts the informed player's offer. This means that the informed player is able to realize the concealing demand even if he is weak. Especially in the game playing experiment we observed the "exploitation" of a revealing weak player. Opponents of both types were able to reach considerably high payoffs against revealers, which is in accord with the sequential equilibrium.

Although the sequential equilibrium cannot explain the observed behavior of strong players, aspects of the behavior of the weak players can be explained by a reasoning in the spirit of the sequential equilibrium strategy selected by Chatterjee and Samuelson (1988).

APPENDIX A.   INSTRUCTIONS OF THE GAME PLAYING EXPERIMENT

The rules of the two-person bargaining game with incomplete information were described to the subjects orally from a prepared text. This introduction lasted about 20 minutes. Firstly, a sheet of paper was distributed to all subjects, the *note on the two-person bargaining game*, containing a summary of the rules discussed in the introduction (a translation from the German original into English is given in figure A.1). Furthermore, the name of the subject and the number of the cubicle in the laboratory where he was seated after the introduction was provided on this sheet. On an extra sheet the subject's computer screen was displayed in order make the subjects familiar with the screen layout and the operation of the software (see figure A.2). The prepared introduction (for $\alpha = 45$) proceeded as follows.

Organization:
You are going to participate in a two-person bargaining experiment. Two participants can allocate a fixed amount of money (called the coalition value), in case they agree on the allocation among themselves. If they fail to agree, each participant receives a predetermined alternative value.
You participate in 16 plays.
The bargaining proceeds via the computer network. No verbal communication is possible.
Your bargaining partner is another participant of this session. He/she remains anonymous and will change from one play to another. But, you may meet the same partner again.

Values and Information:
The coalition value and the alternative value are displayed in the 'payoff window'. The coalition value is always 100.
Each participant knows his/her own alternative value but is ignorant about the alternative value of the bargaining partner. The participant remains ignorant of the partner's alternative value, even after the termination of the play.
The alternative values of both partner are chosen by random before the beginning of each play. They can be either 0 or 45, both equally likely (probability ½).
The random draws are independent of previous draws and the random draw of one partner

is independent of the random draw of the other bargaining partner.

Therefore four constellations of the alternative values are possible: you have 0 and your partner has 0; you have 0 and your partner has 45; you have 45 and your partner has 0; you have 45 and your partner has 45.

But remember, you only see your alternative value and not the alternative value of the bargaining partner.

The alternative value of your bargaining partner will not be made known to you after the end of the bargaining.

Course of the bargaining:

One partner is randomly chosen at the beginning of the bargaining. He/she has the options:
- *break off. The play ends. Each partner receives his/her alternative value.*
- *demand an integer value from [0,100]. The play continues with the bargaining partner.*

The partner has the following options:
- *accept the other partner's demand. The game ends. He/she receives 100−demand as outcome and the partner receives his/her demand.*
- *break off. The play ends. Each partner receives his/her alternative value.*
- *demand an integer value from [0,100]. The play continues with the other partner.*

The play proceeds until either an agreement is reached or one partner breaks off.

There is no time limit and no limit to the number of bargaining steps.

Discounting:

With each proposal, after the initial one, the coalition value and the alternative values of both partners are multiplied by .99 (discounted by 1%).

You bargain over the allocation of the 100, but in case of agreement or break off you receive the discounted values.

At the end of each play you will explicitly be informed about your payoff.

Point-to-cash rate:

Immediately after the session each point you received is rewarded with 6 Pfennig.

Your objective should be to maximize your payoff!

Name: _____   Terminal No.: ____

Note on the Two-Person Bargaining Experiment

Each participant participates in 16 different plays of two-person bargaining.

Bargaining situation:

The coalition value is always 100. Before the bargaining starts, the alternative values of the two bargaining partners are randomly determined. For both partners the alternative is with equal probability either 0 or 45. The random choice of your alternative value is independent of the random choice of the alternative value of your bargaining partner. Furthermore, these choices are independent of previous choices.

Each partner knows his/her own alternative value, but not the one of the bargaining partner. The alternative value of the partner is not made known after the end of the bargaining.

Options of a decider:

ACCEPT   A bargaining partner accepts the proposal of the other partner. The bargaining ends with the proposed allocation as outcomes.

PROPOSE   A bargaining partner proposes an allocation of the coalition value.

BREAK OFF   A bargaining partner breaks off the bargaining and both partners receive their alternative values as outcomes.

One bargaining partner is randomly chosen to be the first decider (both equally likely).

Discount:

With each proposal after the initial one, the coalition value as well as both alternative values are discounted by 1%. The gain from a bargaining is the discounted outcome.

Point-to-cash rate:

Immediately after the experiment, each point is rewarded with 6 Pfennig.

Figure A.1:   Note on the Two-Person Bargaining Experiment

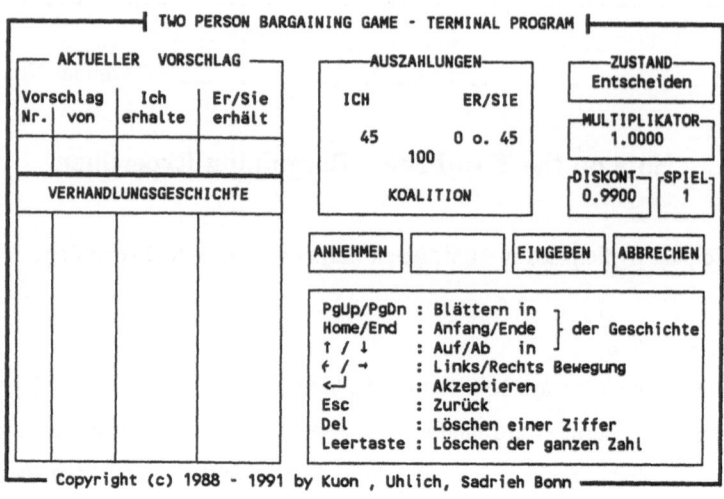

Figure A.2:   The subject's computer screen

Translation of the key words into English:

Abbrechen	-	*break off*
Aktueller Vorschlag	-	*pending proposal*
Annehmen	-	*accept*
Auszahlungen	-	*payoffs*
Diskont	-	*discount*
Eingeben	-	*propose*
Entscheiden	-	*decide*
Er/Sie	-	*he/she*
Er/Sie erhält	-	*he/she receives*
Ich	-	*I*
Ich erhalte	-	*I receive*
Koalition	-	*coalition*
Multiplikator	-	*aggregated discount factor*
Spiel	-	*play*
Verhandlungsgeschichte	-	*history of the bargaining*
Vorschlag .. von	-	*proposal .. by*
Zustand	-	*state*
0 o. 45	-	*0 or 45*

The explanation of the options of the bargaining partners were supported by an example at the blackboard. It showed a picture of the computer screen and the options and their consequences were demonstrated with a carefully designed example in order to avoid biasing the results. In the example the coalition value was 317, the proposer had the alternative value 98 and proposes the allocation (103,214). The values were chosen such that they are neither prominent with respect to the number nor with respect to the allocation.

After each play a payoff information window was displayed on the subject's screen. The following figure displays the translation of the screen into English for the example that the subject had an outcome of 53 after 26 steps.

Payoff Window

Your bargaining outcome is 53 after 26 steps at a discount of .778.

YOUR PAYOFF is $53 \cdot .778 = 41.22$

Figure A.3: Payoff window

After the introduction questions concerning the rules were answered and finally all subjects were seated in the laboratory at the preassigned places.

APPENDIX B.   STATISTICAL TESTS

This appendix gives an overview over the statistical tests that are used in the book. Their applicability to the actual problem as well as a brief procedural description is given. For a more detailed description, see Siegel (1956). A concise mathematical discussion can be found in Hájek (1969).

For each parameter value of α (treatment) six independent sessions were conducted, which form six statistically independent observations (*subject groups*). The observations in between a session are matched, since the players were in interaction with each other.

FRIEDMAN TWO-WAY ANALYSIS OF VARIANCE

The *Friedman two-way analysis of variance* is applied to test whether there is a significant difference in, for example the outcome, among the different experience levels of the players. For each subject group the outcome in one experience level is calculated as the sum over each subject's outcome over all of his 4 plays of the experience level. Thus, this is a sum over 24 ($=$ 6 players $\cdot$ 4 plays) dependent values. Notice, that these sums are comparable among the experience levels, since the same games were played in each experience level. The outcome sums of the different experience levels are matched samples (they are caused by the same players) and it should be tested whether they differ significantly.

The Friedman two-way analysis of variance is applicable to situations where N independent subject groups are observed under K conditions. It tests whether there is a difference in the performance of the conditions. The *null hypothesis* is that there are no significant differences between the K matched samples.

H_0:   The K samples have been drawn from the same population.

H_1:   The K samples have been drawn from different populations.

The Friedman two-way analysis of variance is a two-sided test. The direction of the deviation is not specified in the alternative.

By the example of the outcomes for $\alpha = 30$, the course of the test will be demonstrated. The following table gives the outcome sum over the 24 plays for each independent subject group

$(G_1, \ldots, G_6)$ and for each experience level $(L_1, \ldots, L_4)$. This is a table of $N=6$ rows and $K=4$ columns. The rows represent the groups and the columns represent the experience levels.

Table A.1:   Outcome sums for $\alpha=30$

$\alpha=30$	L_1	L_2	L_3	L_4
G_1	980	1120	1090	980
G_2	1160	1130	1090	1200
G_3	1090	980	880	1130
G_4	1020	1160	1200	1200
G_5	1200	1160	1160	1130
G_6	980	1200	1120	1200

The values of each row have to be ranked, and for each column j the rank sum R_j has to be calculated.

Table A.2:   Ranks of the outcome sums for $\alpha=30$

$\alpha=30$	L_1	L_2	L_3	L_4
G_1	3.5	1	2	3.5
G_2	2	3	4	1
G_3	2	3	4	1
G_4	4	3	1.5	1.5
G_5	1	2.5	2.5	4
G_6	4	1.5	3	1.5
R_j	16.5	14	17	12.5

Under the assumption of the null hypothesis it has to be expected that the ranks 1 to 4 occur with equal frequency in each column, this means, the values R_j are expected to be equal. Whether there is a significant difference in the rank sums of the columns is decided by the

Friedman two-way analysis of variance with the following test statistic.

$$\chi_r^2 = \frac{12}{NK(K+1)} \sum_{j=1}^{K} R_j^2 - 3N(K+1)$$

For N=6 and K=4 it reduces to:

$$\chi_r^2 = \frac{1}{10} \sum_{j=1}^{4} R_j^2 - 90.$$

The distribution of χ_r^2 is approximated by the χ^2-distribution with $K-1$ degrees of freedom. If, for a given significance level, the calculated value of χ_r^2 exceeds the χ^2-value with $K-1$ degrees of freedom, the null hypothesis has the be rejected in favor of the alternative. But, notice that this is only true if N and K are not too small. For our application of N=6 and K=4, the χ^2-distribution can be used.

For the considered example $\chi_r^2 = 1.35$. At a level of .01 it is not possible to reject the null hypothesis.

If the null hypothesis is rejected by the Friedman two-way analysis of variance, the alternative of this two-sided test does neither specify which of the K samples are different, nor does it specify the direction of the difference. In this case the K samples can be compared pairwise by the Wilcoxon matched-pairs signed-ranks test. This is a test for two matched samples, which, in its one-sided version, specifies the direction of the difference in the alternative.

WILCOXON MATCHED-PAIRS SIGNED-RANKS TEST

A test of the difference of two matched samples is the *Wilcoxon matched-pairs signed-ranks test*. It is applicable if pairs of matched variables are observed for each independent subject group. The *null hypothesis* is that there is no difference between the two samples. In the two-sided version of the Wilcoxon matched-pairs signed-ranks test, the alternative does not predict the direction, while in the one-sided version the direction of the difference has to be specified a priori.

H_0:   The two samples have been drawn from the same population.

H_1:   (two-sided) The two samples have been drawn from different population.

The course of the test shall be demonstrated by the example of the verification of the difference between the first demand of the first mover and the first demand of the second mover. These are two matched observations which can be observed in each play. For each subject group the sum of the 48 (= 4 experience levels · 4 plays per experience levels · 3 parallel plays of the 6 players) initial demands of the first mover and the sum of the 48 initial demands of the second mover is calculated. These two values form the matched pair for every independent subject group. Notice, that these values are comparable since each subject group played according to the same experimental setup.

The following table gives the sum of the initial demands of the first mover S_1 and the sum of the initial demands of the second mover S_2. The difference $S_1 - S_2$ is computed and a rank is assigned to its absolute value. This rank receives the sign of the difference.

Table A.3:   Sum of the initial demands for the first and the second mover for $\alpha = 30$

$\alpha = 30$	S_1	S_2	$S_1 - S_2$	Signed rank
G_1	3399	3158	241	6
G_2	2805	2645	160	5
G_3	3019	2955	64	3
G_4	2745	2623	122	4
G_5	2565	2553	12	1
G_6	2853	2873	−20	−2

Under the assumption of the null hypothesis it has to be expected that half of the high ranks as well as half of the low ranks have a negative sign and a positive sign, respectively. Let T_- be the sum of the ranks with a negative sign, T_+ be the sum of the ranks with a positive sign, and $T = \min(T_-, T_+)$. If, for a given significance level, T is smaller or equal to the critical value (see a table for the Wilcoxon matched-pairs signed-ranks test) then the null hypothesis has to be rejected in favor of the alternative.

In a one-sided test, the alternative has to specify whether $T=T_-$ or $T=T_+$, this means whether the minimal sum is the sum of the negative ranks or the sum of the positive ranks. For example:

H_1: (one-sided) $T=T_-$, the first sample is greater than the second sample.

Since we are interested in the direction of a possible difference we shall always work with the one-sided test.

In the example of the comparison of the initial demands, we have $T_-=2$ and $T_+=19$, hence $T=T_-=2$. The critical value of the Wilcoxon matched-pairs signed-ranks test for 6 observations is $T=0$. This corresponds to a significance level of .025 one-sided and a level of .05 two-sided. In the example it is not possible to reject the hypothesis.

Accordingly, in our framework of 6 independent observation we can only reject the hypothesis of non-difference, if all differences have the same sign.

THE ORDER TEST

The *order test* is designed to test whether a sequence of observations follows a trend. We shall apply this test for data which are measured on the four experience levels. Consider for example the initial demand of the first mover. In order to examine whether the initial demands, for example, fall with the experience or do not follow a trend the order test can be applied. Therefore the sum of the initial demands is calculated on each experience level and ranks are assigned to these sums (assign, without loss of generality, the rank 1 to the greatest sum). If these values follow a perfect decreasing trend the rank order has to be 1 2 3 4. A measure of the "difference from the perfect order" is the number of *inversions*. This is the number of pairwise changes that has to be performed in order to transform the given order into the order 1 2 3 4. Let us illustrate this procedure with the concrete example of the initial demands for $\alpha=70$.

For each of the six independent subject groups the sum of the initial demands in each experience level was calculated and ranks were assigned to these values in between each subject group. The number of inversions that has to be performed in order to transform the actual rank order into the order 1 2 3 4 was calculated. The sum of the number of inversions over all subject groups is the test statistic to decide whether a trend is observable. In this example this sum is equal to 23.

Table A.4:   Order test for the initial demands for $\alpha = 70$

Subject group	Rank of the sum of initial demands in level...				Number of inversions
	1	2	3	4	
1	4	2	1	3	4
2	4	3	2	1	6
3	3	4	1	2	4
4	2	4	3	1	4
5	2	1	4	3	2
6	3	1	4	2	3

There are 24 different possibilities to assign four ranks. The *null hypothesis* of the order test is that in each subject group the order of the observed values is arbitrary, which means that all 24 possibilities occur with the same probability.

H_0:   The rank order is arbitrary

The alternative of the two-sided test is:

H_1:   (two-sided) There is an increasing or a decreasing trend in the data

The two possible alternatives of the one-sided tests are:

H_1:   (one-sided) There is an increasing trend in the data

H_1:   (one-sided) There is a decreasing trend in the data

The sixth convolution of the distribution of the inversions among the 24 possibilities allows to identify the values which are likely to be expected as sums of the inversion numbers of six subject groups under the hypothesis, and those which recommend to reject the hypothesis at a given significance level. The following figure shows the density of the sixth convolution of the distributions of the inversions.

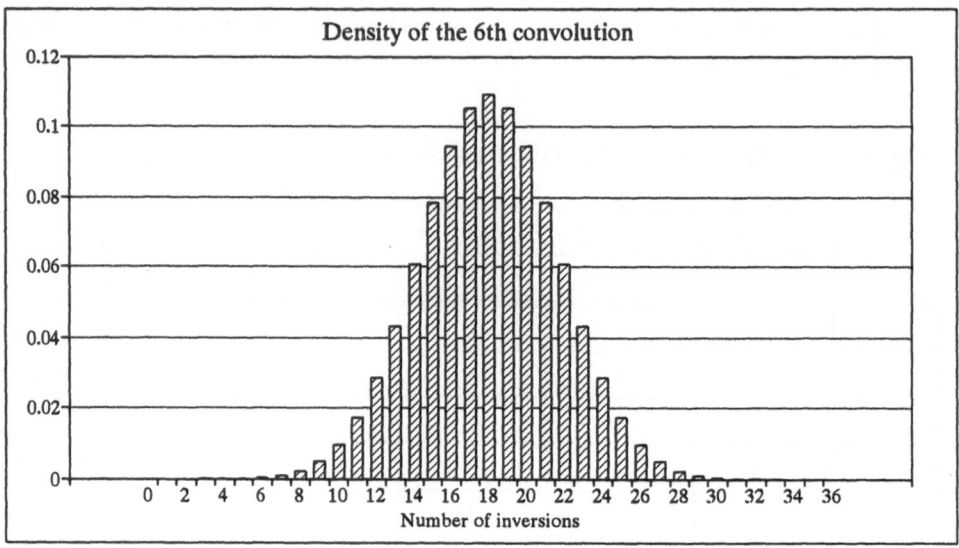

Figure A.4:   Density of the sixth convolution

A value of 23 leads to a rejection of the null hypothesis at a significance level of .107 in an one-sided test in favor of the alternative of an increasing order. A value of 24 leads to a rejection at a level of .06. Usually, we want to achieve a significance level of at most .1. Since the order test with six independent subject groups cannot reach a level of exactly .1, we are sometimes satisfied with rejecting the null hypothesis at a level of .107, which is closest to .1.

A similar test based on the same idea of rank orders was conducted by Selten (1967a).

SPEARMAN RANK CORRELATION ANALYSIS

The *Spearman rank correlation test* determines a coefficient of the correlation between to samples and, moreover, tests whether this correlation is significant. The *null hypothesis* assumes no correlation between the two samples in the population of the subjects.

H_0:   There is no correlation between the two samples.

The two possible alternatives of the one-sided tests are:

H_1:   The two samples are in positive correlation.

H_1:   The two samples are in negative correlation.

The one-sided test has to specify the type of the correlation.

The course of the test will be demonstrated by the example of the correlation between the initial demand and the number of concessions for $\alpha=30$. For each independent subject group the average initial demand (A_1) and the average number of concessions (A_2) is calculated.

Table A.5:   Correlation of the initial demand and the number of concessions for $\alpha=30$

$\alpha=30$	A_1	A_2	Rank A_1	Rank A_2	d_i
G_1	68.29	3.40	6	6	0
G_2	57.99	1.47	2	3	-1
G_3	63.02	1.74	5	4	1
G_4	59.02	0.67	3	2	1
G_5	55.96	0.58	1	1	0
G_6	61.76	2.48	4	5	-1

After the values of each feature are ranked, the Spearman rank correlation coefficient can be computed as follows (d_i is the difference of the ranks of the two features for group i).

$$\rho = 1 - \frac{6}{N^3 - N} \cdot \sum_{i=1}^{N} d_i^2$$

If the two features were in perfect correlation, the difference between the two ranks would always be zero. The Spearman rank correlation coefficient ρ describes the degree of correlation between the two features.

For N=6, this formula reduces to:

$$\rho = 1 - \frac{1}{35} \sum_{i=1}^{N} d_i^2 .$$

In the example the value is $\rho=.886$.

For $N=6$, the critical value is $\rho = .829$ at a significance level of .05 (one-sided), and $\rho = .943$ at a significance level of .01 (one-sided). If the calculated ρ is greater or equal to the critical value, the null hypothesis has to be rejected in favor of the alternative.

THE BINOMIAL TEST

The *Binomial test* can be applied if the observations of a sample can be divided into two disjoint classes A and B. Let p be the (theoretical) probability of observations in class A. Accordingly $1-p$ is the (theoretical) probability of observations in class B. The Binomial test is designed to test whether the frequencies of observations in the two classes deviate significantly from what can be expected if the theoretical probabilities are p and $1-p$. We shall apply the test for $p=\frac{1}{2}$, this means we shall state the null hypothesis that the observations are equally likely in both classes.

H_0: $p=1-p=\frac{1}{2}$.

The two possible alternatives of the one-sided tests are:

H_1: $p>\frac{1}{2}$.

H_1: $p<\frac{1}{2}$.

The alternative of the two-sided test is:

H_1: $p \neq 1-p$.

Consider a one-sided Binomial test with $N=6$ observations and an alternative hypothesis which specifies the lower probability in class A. At a significance level of .109 the null hypothesis has to be rejected if at most one observation is in class A, and at a significance level of .016 the null hypothesis has to be rejected only if no observation is in class A. Usually, we want to achieve a significance level of at most .1. Since the Binomial test with six observations cannot reach a level of exactly .1, we are sometimes satisfied with rejecting the null hypothesis at a level of .109, which is closest to .1.

REFERENCES

ALBERS, W. AND G. ALBERS (1983), "On the Prominence Structure of the Decimal System", in: R.W. Scholz (ed.), *Decision Making under Uncertainty*, Elsevier Science Publishers B.V., North Holland, 271-287.

AUSUBEL, L.M. AND R.J. DENECKERE (1989), "A Direct Mechanism Characterization of Sequential Bargaining with One-Sided Incomplete Information", *Journal of Economic Theory*, 48, 18-46.

BARTOS, O.J. (1974), *The Process and Outcome of Negotiations*, New York.

BARTOS, O.J. (1978), "Negotiation and Justice", in: H. Sauermann (ed.), *Contributions to Experimental Economics Vol. 7*, J.C.B. Mohr, Tübingen, 103-126.

BARTOS, O.J., TIETZ, R. AND C. MCLEAN (1983), "Toughness and Fairness in Negotiations", in: R. Tietz (ed.), *Aspiration Levels in Bargaining and Decision Making*, Springer Lecture Notes in Economics and Mathematical Systems, Berlin, Heidelberg, New-York, Tokyo, 35-51.

BINMORE, K., SHAKED, A. AND J. SUTTON (1985), "Testing Noncooperative Bargaining Theory: A Preliminary Study", *American Economic Review*, 75, 1178-1180.

BOLTON, G.E. (1991), "A Comparative Model of Bargaining: Theory and Evidence", *American Economic Review*, 81, 1096-1136.

CHATTERJEE, K. AND L. SAMUELSON (1987), "Bargaining with Two-Sided Incomplete Information: An Infinite Horizon Model with Alternating Offers", *Review of Economic Studies*, 54, 175-192.

CHATTERJEE, K. AND L. SAMUELSON (1988), "Bargaining under Incomplete Information: The Unrestricted Offer Case", *Operations Research*, 36, 605-618.

CROSON, R.T.A. (1992), "Information in Ultimatum Games: An Experimental Study", *mimeo*.

CROSS, J.G. (1965), "A Theory of the Bargaining Process", *American Economic Review*, 67-94.

CROTT, H.W., MÜLLER, G.F. AND P.L. HAMEL (1978), "The Influence of the Aspiration Level, of the Level of Information and Bargaining Experience on the Process and Outcome in a Bargaining Situation", in: H. Sauermann (ed.), *Contributions to Experimental Economics Vol. 7*, J.C.B. Mohr, Tübingen, 211-230.

FORSYTHE, R., KENNAN, J. AND B. SOPHER (1991), "An Experimental Analysis of Strikes in Bargaining Games with One-Sided Private Information", *American Economic Review*, 81, 253-278.

FROHLICH, N. AND J. OPPENHEIMER (1984), "Beyond Economic Man: Altruism, Egalitarism and Difference Maximizing", *Journal of Conflict Resolution*, 28, 3-24.

GAMSON, W.A. (1961), "A Theory of Coalition Formation", *American Sociological Review*, 26, 373-382.

GÜTH, W., SCHMITTENBERGER, R. AND B. SCHWARZ (1982), "An Experimental Analysis of Ultimatum Bargaining", *Journal of Economic Behavior and Organization*, 3, 367-388.

GÜTH, W. AND R. TIETZ (1988), "Ultimatum Bargaining for a Shrinking Pie: An Experimental Analysis", in: R. Tietz, W. Albers, and R. Selten (eds.), *Bounded Rational Behavior in Experimental Games and Markets*, Proceedings of the Fourth Conference on Experimental Economics, New York, Springer Verlag, 111-128.

HÁJEK, J. (1969), *Nonparametric Statistics*, Holden-Day, San Francisco, Cambridge, London, Amsterdam.

HARSANYI, J.C. (1956), "Approaches to the Bargaining Problem Before and After the Theory of Games: A Critical Discussion of Zeuthen's, Hick's, and Nash's Theories, *Econometrica*, 24, 144-157.

HARSANYI, J.C. (1967-1968), "Games with Incomplete Information played by 'Bayesian' Players", Parts I-III, *Management Science*, 14, 158-182, 320-334, 486-502.

HARSANYI, J.C. AND R. SELTEN (1972), "A Generalized Nash Solution for Two-Person Bargaining Games with Incomplete Information", *Management Science*, 18, P80-P106.

HOFBAUER, J. AND K. SIGMUND (1988), *The Theory of Evolution and Dynamical Systems*, Cambridge University Press, Cambridge.

HOFFMAN, A. AND M. SPITZER (1982), "The Coase Theorem: Some Experimental Tests", *Journal of Law and Economics*, 25, 73-98.

HOFFMAN, A. AND M. SPITZER (1985), "Entitlements, Rights and Fairness: An Experimental Examination of Subjects Concepts of Distributive Justice", *Journal of Legal Studies*, 15, 254-297.

HOFFMAN, A., MCCABE, K., SHACHAT K. AND V.L. SMITH (1992), "Preferences, Property Rights and Anonymity in Bargaining Games", *Discussion Paper 92-8, Department of Economics, University of Arizona*.

HOGGATT, A.C., SELTEN, R., CROCKETT, D., GILL, S. AND J. MOORE (1978), "Bargaining Experiments with Incomplete Information", in: H. Sauermann (ed.), *Contributions to Experimental Economics Vol. 7*, J.C.B. Mohr, Tübingen, 127-178.

KACHELMAIER, S.J., LIMBERG, S.T. AND M.S. SCHADEWALD (1991), "Fairness in Markets: A Laboratory Investigation", *Journal of Economic Psychology*, 12, 447-464.

KAHNEMAN, D., KNETSCH, J. AND R. THALER (1986a), "Fairness and the Assumptions of Economics", *Journal of Business*, 59, S285-S300.

KAHNEMAN, D., KNETSCH, J. AND R. THALER (1986b), "Fairness as a Constraint on Profit Seeking: Entitlements in the Market", *American Economic Review*, 76, 728-741.

KENNAN, J. AND R. WILSON (1993), "Bargaining with Private Information", *Journal of Economic Literature*, XXXI, 45-104.

KESER, C. (1992), "Experimental Duopoly Markets with Demand Inertia: Game-Playing Experiments and the Strategy Method", Springer Lecture Notes in Economics and Mathematical Systems No. 391, Berlin, Heidelberg, New York, Tokyo.

KUON, B. (1991), "Typical Behavior in the German Election Markets", *Discussion Paper B-204, University of Bonn*.

KUON, B. (1993), "Measuring the Typicalness of Behavior", *Mathematical Social Sciences*, 26, 35-49.

KUON, B. AND G.R. UHLICH (1993), "The Negotiation Agreement Area: An Experimental Analysis of Two-Person Characteristic Function Games", *Group Decision and Negotiation*, 2, 323-345.

MAYNARD SMITH, J. AND G.R. PRICE (1973), "The Logic of Animal Conflict", *Nature*, 246, 15-18.

MITZKEWITZ, M. AND R. NAGEL (1993), "Experimental Results on Ultimatum Games with Incomplete Information", *International Journal of Game Theory*, 22, 171-198.

MYERSON, R. (1979), "Incentive Compatibility and the Bargaining Problem", *Econometrica*, 47, 61-73.

MYERSON, R. (1984), "Two-Person Bargaining Problems with Incomplete Information", *Econometrica*, 52, 461-487.

NASH, J.F. (1950), "The Bargaining Problem", *Econometrica*, 18, 155-162.

OCHS, J. AND A.E. ROTH (1989), "An Experimental Study of Sequential Bargaining", *American Economic Review*, 79, 335-384.

PRASNIKAR, V. AND A.E. ROTH (1992), "Considerations of Fairness and Strategy: Experimental Data from Sequential Games", *Quarterly Journal of Economics*, 865-888.

ROTH, A.E. AND M.W.K. MALOUF (1979), "Game-Theoretic Models and the Role of Information in Bargaining", *Psychological Review*, 86, 574-594.

ROTH, A.E., M.W.K. MALOUF AND J.K. MURNIGHAN (1981), "Sociological versus Strategic Factors in Bargaining", *Journal of Economic Behavior and Organization*, 2, 153-177.

ROTH, A.E. AND J.K. MURNIGHAN (1982), "The Role of Information in Bargaining: An Experimental Study", *Econometrica*, 50, 1123-1142.

RUBINSTEIN, A. (1982), "Perfect Equilibrium in a Bargaining Model", *Econometrica*, 50, 97-109.

RUBINSTEIN, A. (1985a), "A Bargaining Model with Incomplete Information about Time Preferences", *Econometrica*, 53, 1151-1172.

RUBINSTEIN, A. (1985b), "Choice of Conjectures in a Bargaining Game with Incomplete Information", in: A.E. Roth (ed.), *Game Theoretic Models of Bargaining*, Cambridge University Press, Cambridge, 99-114.

SCHELLING, A. (1960), "The Strategy of Conflict", *Harvard University Press*, Cambridge.

SCHUSTER, P., SIGMUND, K., HOFBAUER, J. AND R. WOLFF (1981), "Selfregulation of Behavior in Animal Societies. Part I: Symmetric Contests, Part II: Games between two Populations with Selfinterest", *Biological Cybernetics*, 40, 17-25.

SELTEN, R. (1967a), "Ein Oligopolexperiment mit Preisvariation und Investition", in: H. Sauermann (ed.), *Beiträge zur experimentellen Wirtschaftsforschung*, J.C.B. Mohr, Tübingen, 103-135.

SELTEN, R. (1967b), "Die Strategiemethode zur Erforschung des eingeschränkt rationalen Verhaltens im Rahmen eines Oligopolexperiments", in: H. Sauermann (ed.), *Beiträge zur experimentellen Wirtschaftsforschung*, J.C.B. Mohr, Tübingen, 136-168.

SELTEN, R. (1975), "Bargaining under Incomplete Information - A Numerical Example", in: Becker, O. and R. Richter (eds.), *Dynamische Wirtschaftsanalyse*, J.C.B. Mohr, Tübingen, 203-232.

SELTEN, R. (1980),"A Note on Evolutionarily Stable Strategies in Asymmetric Animal Conflicts", *Journal of theoretical Biology*, 84, 93-101.

SELTEN, R. (1982), "Einführung in die Theorie der Spiele mit unvollständiger Information", in: *Schriften des Vereins für Socialpolitik*, Gesellschaft für Wirtschafts- und Sozialwissenschaften, Neue Folge Band 126, Information in der Wirtschaft, 81-147.

SELTEN, R. (1983), "Evolutionary Stability in Extensive Two-Person Games", *Mathematical Social Sciences*, 5, 269-363.

SELTEN, R. (1987), "Equity and Coalition Bargaining in Experimental Three-Person Games", in: A.E. Roth (ed.), *Laboratory Experimentation in Economics*, Cambridge University Press.

SELTEN, R. (1988), "Evolutionary Stability in Extensive Two-Person Games - Correction and Further Development", *Mathematical Social Sciences*, 16, 223-266.

SELTEN, R. (1991), "Properties of a Measure of Predictive Success", *Mathematical Social Sciences*, 21, 153-167.

SELTEN, R. AND W. KRISCHKER (1983), "Comparison of Two Theories for Characteristic Function Experiments", in: R. Tietz (ed.), *Aspiration Levels in Bargaining and Economic Decision Making*, Springer Lecture Notes in Economics and Mathematical Systems, Berlin-Heidelberg-New York-Tokyo, 259-264.

SELTEN, R. AND B. KUON (1993), "Demand Commitment Bargaining in Three-Person Quota Game Experiments", *International Journal of Game Theory*, 22, 261-277.

SELTEN, R., M. MITZKEWITZ AND G.R. UHLICH (1988), "Duopoly Strategies Programmed by experienced Players", *Discussion Paper B-106, University of Bonn*.

SIEGEL, S. (1956), *Nonparametric Statistics for the Behavioral Sciences*, McGraw-Hill, New York, Toronto, London.

SIEGEL, S. AND L.E. FOURAKER (1960), *Bargaining and Group Decision Making*, McGraw-Hill, New York.

THALER, R.H. (1988), "Anomalies: The Ultimatum Game", *Journal of Economic Perspectives*, 2, 195-206.

TIETZ, R. (1978), "Entscheidungsprinzipien der bilateralen Anspruchsanpassung", in: E. Helmstädter (ed.), *Neuere Entwicklungen in den Wirtschaftswissenschaften*, SdVSp., N.F. 98, Berlin, 431-453.

TIETZ, R. (1984), "The Prominence Standard", Part 1, *Discussion Paper A 18*, Frankfurt.

TIETZ, R. AND H.J. WEBER (1972), "On the Nature of the Bargaining Process in the Kresko-Game", in: H. Sauermann (ed.), *Contributions to Experimental Economics Vol. 3*, J.C.B. Mohr, Tübingen, 305-334.

TIETZ, R., WEBER, H.J., VIDMAJER, U. AND C. WENTZEL (1978), "On Aspiration-Forming Behavior in Repetitive Negotiations", in: H. Sauermann (ed.), *Contributions to Experimental Economics Vol. 7*, J.C.B. Mohr, Tübingen, 88-102.

UHLICH, G.R. (1990), *Descriptive Theories of Bargaining: An Experimental Analysis of Two- and Three-Person Characteristic Function Bargaining*, Springer Lecture Notes in Economics and Mathematical Systems No. 341, Berlin, Heidelberg, New York, Tokyo.

VAN DAMME, E.E.C. (1987), *Stability and Perfection of Nash Equilibria*, Springer Verlag, Berlin, Heidelberg, New York, London, Paris, Tokyo.

YAARI, M.E. AND M. BAR-HILLEL (1984), "On Dividing Justly", *Social Choice and Welfare*, 1, 1-24.

YUKL, G. (1974), "Effects of the Opponent's Initial Offer, Concession Magnitude, and Concession Frequency on Bargaining Behavior", *Journal of Personality and Social Psychology*, 30, 323-335.

ZEUTHEN, F. (1930), *Problems of Monopoly and Economic Warfare*, London.

Lecture Notes in Economics and Mathematical Systems

For information about Vols. 1–234
please contact your bookseller or Springer-Verlag

Vol. 367: M. Grauer, D. B. Pressmar (Eds.), Parallel Computing and Mathematical Optimization. Proceedings. V, 208 pages. 1991.

Vol. 368: M. Fedrizzi, J. Kacprzyk, M. Roubens (Eds.), Interactive Fuzzy Optimization. VII, 216 pages. 1991.

Vol. 369: R. Koblo, The Visible Hand. VIII, 131 pages.1991.

Vol. 370: M. J. Beckmann, M. N. Gopalan, R. Subramanian (Eds.), Stochastic Processes and their Applications. Proceedings, 1990. XLI, 292 pages. 1991.

Vol. 371: A. Schmutzler, Flexibility and Adjustment to Information in Sequential Decision Problems. VIII, 198 pages. 1991.

Vol. 372: J. Esteban, The Social Viability of Money. X, 202 pages. 1991.

Vol. 373: A. Billot, Economic Theory of Fuzzy Equilibria. XIII, 164 pages. 1992.

Vol. 374: G. Pflug, U. Dieter (Eds.), Simulation and Optimization. Proceedings, 1990. X, 162 pages. 1992.

Vol. 375: S.-J. Chen, Ch.-L. Hwang, Fuzzy Multiple Attribute Decision Making. XII, 536 pages. 1992.

Vol. 376: K.-H. Jöckel, G. Rothe, W. Sendler (Eds.), Bootstrapping and Related Techniques. Proceedings, 1990. VIII, 247 pages. 1992.

Vol. 377: A. Villar, Operator Theorems with Applications to Distributive Problems and Equilibrium Models. XVI, 160 pages. 1992.

Vol. 378: W. Krabs, J. Zowe (Eds.), Modern Methods of Optimization. Proceedings, 1990. VIII, 348 pages. 1992.

Vol. 379: K. Marti (Ed.), Stochastic Optimization. Proceedings, 1990. VII, 182 pages. 1992.

Vol. 380: J. Odelstad, Invariance and Structural Dependence. XII, 245 pages. 1992.

Vol. 381: C. Giannini, Topics in Structural VAR Econometrics. XI, 131 pages. 1992.

Vol. 382: W. Oettli, D. Pallaschke (Eds.), Advances in Optimization. Proceedings, 1991. X, 527 pages. 1992.

Vol. 383: J. Vartiainen, Capital Accumulation in a Corporatist Economy. VII, 177 pages. 1992.

Vol. 384: A. Martina, Lectures on the Economic Theory of Taxation. XII, 313 pages. 1992.

Vol. 385: J. Gardeazabal, M. Regúlez, The Monetary Model of Exchange Rates and Cointegration. X, 194 pages. 1992.

Vol. 386: M. Desrochers, J.-M. Rousseau (Eds.), Computer-Aided Transit Scheduling. Proceedings, 1990. XIII, 432 pages. 1992.

Vol. 387: W. Gaertner, M. Klemisch-Ahlert, Social Choice and Bargaining Perspectives on Distributive Justice. VIII, 131 pages. 1992.

Vol. 388: D. Bartmann, M. J. Beckmann, Inventory Control. XV, 252 pages. 1992.

Vol. 389: B. Dutta, D. Mookherjee, T. Parthasarathy, T. Raghavan, D. Ray, S. Tijs (Eds.), Game Theory and Economic Applications. Proceedings, 1990. IX, 454 pages. 1992.

Vol. 390: G. Sorger, Minimum Impatience Theorem for Recursive Economic Models. X, 162 pages. 1992.

Vol. 391: C. Keser, Experimental Duopoly Markets with Demand Inertia. X, 150 pages. 1992.

Vol. 392: K. Frauendorfer, Stochastic Two-Stage Programming. VIII, 228 pages. 1992.

Vol. 393: B. Lucke, Price Stabilization on World Agricultural Markets. XI, 274 pages. 1992.

Vol. 394: Y.-J. Lai, C.-L. Hwang, Fuzzy Mathematical Programming. XIII, 301 pages. 1992.

Vol. 395: G. Haag, U. Mueller, K. G. Troitzsch (Eds.), Economic Evolution and Demographic Change. XVI, 409 pages. 1992.

Vol. 396: R. V. V. Vidal (Ed.), Applied Simulated Annealing. VIII, 358 pages. 1992.

Vol. 397: J. Wessels, A. P. Wierzbicki (Eds.), User-Oriented Methodology and Techniques of Decision Analysis and Support. Proceedings, 1991. XII, 295 pages. 1993.

Vol. 398: J.-P. Urbain, Exogeneity in Error Correction Models. XI, 189 pages. 1993.

Vol. 399: F. Gori, L. Geronazzo, M. Galeotti (Eds.), Nonlinear Dynamics in Economics and Social Sciences. Proceedings, 1991. VIII, 367 pages. 1993.

Vol. 400: H. Tanizaki, Nonlinear Filters. XII, 203 pages. 1993.

Vol. 401: K. Mosler, M. Scarsini, Stochastic Orders and Applications. V, 379 pages. 1993.

Vol. 402: A. van den Elzen, Adjustment Processes for Exchange Economies and Noncooperative Games. VII, 146 pages. 1993.

Vol. 403: G. Brennscheidt, Predictive Behavior. VI, 227 pages. 1993.

Vol. 404: Y.-J. Lai, Ch.-L. Hwang, Fuzzy Multiple Objective Decision Making. XIV, 475 pages. 1994.

Vol. 405: S. Komlósi, T. Rapcsák, S. Schaible (Eds.), Generalized Convexity. Proceedings, 1992. VIII, 404 pages. 1994.

Vol. 406: N. M. Hung, N. V. Quyen, Dynamic Timing Decisions Under Uncertainty. X, 194 pages. 1994.

Vol. 407: M. Ooms, Empirical Vector Autoregressive Modeling. XIII, 380 pages. 1994.

Vol. 408: K. Haase, Lotsizing and Scheduling for Production Planning. VIII, 118 pages. 1994.

Vol. 409: A. Sprecher, Resource-Constrained Project Scheduling. XII, 142 pages. 1994.

Vol. 410: R. Winkelmann, Count Data Models. XI, 213 pages. 1994.

Vol. 411: S. Dauzère-Péres, J.-B. Lasserre, An Integrated Approach in Production Planning and Scheduling. XVI, 137 pages. 1994.

Vol. 412: B. Kuon, Two-Person Bargaining Experiments with Incomplete Information. IX, 293 pages. 1994.